50-11

D1395401

AN INTRODUCTION TO COMPUTER SIMULATION IN APPLIED SCIENCE

AN INTRODUCTION TO COMPUTER SIMULATION IN APPLIED SCIENCE

EDITED BY

FARID F. ABRAHAM

IBM Scientific Center
Palo Alto, California

and

Consulting Associate Professor
Materials Science Department
Stanford University
Stanford, California

AND

WILLIAM A. TILLER

Professor
Materials Science Department
Stanford University
Stanford, California

℗ PLENUM PRESS · NEW YORK–LONDON · 1972

Library of Congress Catalog Card Number 72-83047
ISBN 0-306-30579-8

THIS BOOK IS DEDICATED BY FARID ABRAHAM
TO HIS FATHER
ANTONY F. ABRAHAM

Preface

This set of lectures is the outgrowth of a new course in the Department of Materials Science at Stanford University. It was taught collectively by the authors of the various sections and represents an attempt to increase the awareness of students in the materials area of computer simulation techniques and potentialities. The topics often ranged far afield from the materials area; however, the total package served the intended purpose of being an initiation into the world of computer simulation and, as such, made a useful first iteration to the intended purpose. The second iteration, which is in process, deals exclusively with the materials area.

The course was designed to teach students a new way to wrestle with "systems" problems in the materials science work area that require the synthesis and interactions of several disciplines of knowledge. This course was a response to the realization that effective handling of real problems, which are essentially systems problems, is one of the most important attributes of a graduate materials scientist. About a third of the course was devoted to the student's selected problem, in the materials area, which he simulated using the digital computer.

The set of lectures begins with Professor Tiller presenting the essential philosophy for dissociating real problems into a system of identifiable and interacting parts and rationalizing the vital importance of computer simulation in this resolution process. The importance of the computer simulation technique to the synthesis of knowledge within the student and to his acquiring the confidence to become a significant problem-solver is stressed. Finally, some examples are given to illustrate the dissociation of a problem area into a system of subroutines with the relevant parameters and variables identified for a particular level of modeling.

Jacob Fromm presents the basic needs for computation of nonlinear fluid flows. Included are considerations of the governing equations and their finite difference representation. The value of the linear stability analysis of the difference equations is emphasized. Finally, an outline of a working program is given along with listings and a test problem solution. A series of results of the numerical program are discussed with the object of demonstrating the versatility of the program and suggesting potential uses in related areas.

Farid Abraham's presentation discusses a "simulation language" that is easy to use, is powerful in solving a large number of differential equations, and is able to solve the algebraic as well as the differential equations. This simulation language is entitled "The System/360 Continuous System Modeling Program (S/360 CSMP)" and does not require the user to be a proficient computer programmer. S/360-CSMP is illustrated by obtaining numerical solutions for some heat diffusion problems.

George White discusses vapor deposition simulation programs developed by use of Monte Carlo methods to describe the molecular processes of condensation, evaporation, and migration on lattices. The principal application is to systems that permit a comparison with Honig's theoretical work, although the simulation methods are easily applied to a variety of other problems. The results of the simulations demonstrate that Honig's treatment is quite accurate and is a substantial improvement over previous treatments. This agreement also serves to build confidence in the use of Monte Carlo methods in simulating molecular dynamics for vapor deposition studies.

Robert Kortzeborn introduces computational theoretical chemistry via the solution of the Schrödinger equation for the hydrogen and helium atoms. The concept of integral poles and approximate techniques that arise in the two electron integrals are discussed. Molecular systems are then considered with emphasis on the theoretical model and its relationship to the real world. A brief explanation of the Hartree-Fock model is presented followed by a detailed research method for computing multicentered, two-electron integrals via transformation techniques. This method is illustrated with the appropriate mathematics and FORTRAN code.

Finally, Harwood Kolsky describes the physical phenomena occurring in the atmosphere and the problems of modeling them for computer analysis. The numerical methods commonly used in general circulation models are described briefly, and the relative advantages are discussed. An analysis of the computer requirements for global weather calculations is developed, and the need is pointed out for very fast computers capable of executing the equivalent of hundreds of millions of instructions per second.

We are indebted to Jacob Fromm, Harwood Kolsky, Robert Kort-zeborn, and George White for giving us a timely review of the present state of the art of digital simulation of applied science problems and for unfailing cooperation in making this manuscript possible. To Ms. Barbara Merrill, we express a special thanks for typing and organizing the manuscript.

FARID F. ABRAHAM
WILLIAM A. TILLER

Contents

Chapter 1

Rationale for Computer Simulation in Materials Science

W. A. Tiller

Materials Science Department
Stanford University
Stanford, California

I. INTRODUCTION

The intent is to lay a philosophic foundation for the need and place of computer simulation in both the technological endeavors of the materials science field and in the educational life of students proceeding toward careers in this area. We shall begin by discussing the differences and similarities exhibited by the materials science field and other specializations of science and technology. The philosophy presented applies very generally to the whole field of applied science so that the concepts and notions developed in the frame of reference of materials science are readily transferable to the broader field.

We shall see that materials science is not a discipline, like solid state physics, but is more like metallurgy grown up to the stage where it extends its interest and experimentation to all materials and, furthermore, attempts a quantitative evaluation of the multivariable, multiparameter problems encountered with them. Thus materials science is not a discipline in its own right but is located where disciplines converge to give balanced understanding about real problems, e.g., it deals with an ensemble of interacting phenomena where the important characteristics of the event are associated with the interactions.

1

Next we proceed to consideration of several functions of a student's research, what one can hope to gain via the simulation-type analysis of "systems" problems, and what are the important attitudes, characteristics, and abilities that we would wish to find in a "typical" materials science student. Finally, a number of materials problems in the crystallization area are discussed as examples to illustrate their "systems" nature.

II. PATTERNS OF SCIENCE

Why are we concerned with science at all? The answer here is fairly simple—man wants to understand the milieu in which he finds himself. He wants to engineer and control as much of his environment as possible in order to sustain, propagate, and enrich his life. Following this lead, science and engineering appear to have two complementary goals: (1) that of science is the reliable prediction of behavior as a function of ever-changing environment and (2) that of engineering is the generation of materials, devices, attitudes, moralities, philosophies, etc. for producing order and expanding human potentialities in this environment (these are personal definitions rather than generally accepted ones).

As to the patterns of science, the time-honored method of inquiry treats a phenomenon under study (which may be the result of a single event or of an ensemble of interacting events) as a black box whose internal characteristics are unknown but are amenable to probing and analysis. Such a situation is illustrated in the upper portion of Fig. 1 in which we apply some input stimulus (I.S.) to the box and determine some output response (O.R.). By correlating the O.R. with the I.S., information is deduced about the most probable behavior of the box for this degree of variation of the stimulus. We then speculate on models that would reproduce such a spectrum of responses and design critical tests to discriminate between acceptable models. With time, man has learned to recognize clustered phenomena and to dissociate them so that isolated phenomena could be probed and modeled in great detail. This discrimination into isolated phenomena has led to the disciplines of physics, chemistry, mathematics, etc.

Our first step toward determining the behavior of the black box in Fig. 1 is to characterize it in the following form:

$$\frac{\text{O.R.}}{\text{I.S.}} = f(\varepsilon_1, \varepsilon_2, \ldots, \varepsilon_i, \ldots, \varepsilon_n; X_1, X_2, \ldots, X_i, \ldots, X_n) \tag{2.1a}$$

$$\approx f'(\varepsilon_1, \ldots, \varepsilon_j; X_1, \ldots, X_k) \quad \varepsilon_j{}^* \leq \varepsilon_j \leq \varepsilon_j{}^{**} \quad X_k{}^* \leq X_k \leq X_k{}^{**} \tag{2.1b}$$

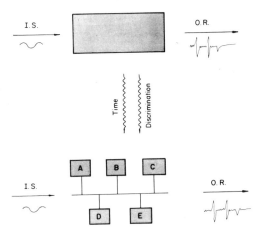

Fig. 1.   Schematic illustration of the "black box" and "systems" approach for studying nature: I.S. = input stimulus; O.R. = output response.

In Eq. (2.1a) f represents the true functional relationship between all the possible material parameters ε_i and the variables X_i of the system, where an unlimited range is allowed for these parameters and variables. In Eq. (2.1b) the observed functional relationship f' between a limited but seemingly sufficient number of parameters and variables is indicated for bounded ranges of the parameters and variables. For the disciplines of mathematics, physics, and chemistry, j and k are generally small and f' is determined to a high degree of accuracy. With the passage of time, our sensing capacity increases so that j and k tend to increase and the bounded ranges of the parameters and variables increase ($\varepsilon_i^{**} - \varepsilon_i^{*}$ increases; $X_i^{**} - X_i^{*}$ increases). One tends to forget that the entire fabric of nature is represented by Eq. (2.1a) and that man has restricted it to conform to Eq. (2.1b) in order to make a manageable assessment of it, i.e., he focuses his attention on one thread of the fabric in order to find reproducibility and thus reliable behavior in this restricted domain. This generates *knowledge* of the thread that is different from *understanding* of the fabric. Some individuals recognize this difference—many do not.

   In a large variety of real situations we encounter in life, the events we wish to understand and control involve the treatment of clustered phenomena that interact strongly with each other and must be considered in association. Such events, which include metallurgy, medicine, technology, business, politics, history, etc., may each be treated as an event conforming to Eq. (2.1); however, we tend to find that j and k are extremely large, many important ε_n and X_n must be neglected, and f' is only poorly defined even for very restricted bounds of the ε_i and X_i. This confines us to a "recipe" or "art" mode of operation in these fields. A preferable characterizing function for such events is illustrated in the lower portion of Fig. 1 and functionally as the following:

$$\frac{\text{O.R.}}{\text{I.S.}} = g(f_1, \ldots, f_g) \tag{2.2a}$$

$$\approx g'(f_1', \ldots, f_j') \tag{2.2b}$$

where the f_i and the f_i' are of the form represented by Eqs. (1.1) and where g and g' represent the exact and observed functional relationships between the various f_i and f_i' respectively. For such associated phenomena, in the analysis or representation, the f_i' can be treated as elemental parts or subsets in the overall system or ensemble and, although the f_i' may be well-characterized for certain disciplines, we must expect the initial reliability of g to be fairly poor.

In order to develop a science of events that conforms to Eq. (2.2) for a substantial range of variation of the ε_i and X_i, we must develop methods of systems analysis for the events. The steps to be taken appear to be (1) the identification of the critical and individual phenomena included in the single black box that encompasses the associated event, i.e., the identification of the f_i in Eqs. (2.2) that are the small boxes A, B, C, D, and E in Fig. 1, (2) the understanding of the f_i in isolation so that a quantitative response spectrum can be determined for a quantitative input stimulus, (3) the understanding of the f_i as they interact with various of the other f_j in pairs, triplets, etc., and (4) the partitioning of the total potential for the associated event into that consumed by the various elemental f_i as they interact with each other and the evaluation of the spectrum of these potentials.

As an example to merely illustrate the point I am trying to make, suppose we want to predict the structure (grain size, shape, degree of macro- and microsegregation, etc.) of a volume v of binary alloy liquid having a solute content C_∞ that is held at some superheated temperature T_0 at time $t = 0$ and then is cooled at its outer surface at a heat extraction rate $\dot{Q}$ per unit area of surface. To gain this information, it is necessary to discriminate at least nine separate f_i as indicated in Table I,[1] at least 20 material parameters enter the description, at least 7 interface variables must be considered that control the processes going on at the interface between the crystals and the liquid, and at least 5 major field equations must be considered. Using the philosophy of Eq. (2.1) would make it rel-

Table I. Crystallization Variables and Parameters[a]

Areas of study	Boundary-value problems	Material parameters	Interface variables	Macroscopic variables	Constraints
Phase equilibria		$\Delta H, T_0, k_0, m_L$		$T_L(C_\infty)$	
Nucleation		$N_0, \Delta T_C$		t	
Solute partitioning	Diffusion equation (C)	D_S, D_L, k_i	$C_i, T_L(C_i), V, S$	C_∞	
Fluid motion	Hydrodynamic equation (u)	ν	δ_C	u_∞	C_∞
Excess solid free energy		$\gamma, \Delta S, \Sigma_i \gamma_i^f, \Sigma_i N_i^f$	T_E		$\dot{T}$
Interface attachment kinetics		μ_1, μ_2			
Heat transport	Heat equation (T)	$K_S, K_L, \alpha_S, \alpha_L$	T_i	T_∞	
Interface morphology	Perturbation response and coupling equations				
Defect generation	Stress equation				

[a] ΔH is the latent heat of fusion, T_0 is the melting temperature of solvent, k_0 is the solute distribution coefficient, m_L is the liquidus slope, N_0 is a parameter related to area of nucleation catalyst surface, ΔT_C is a parameter related to potency of nucleation catalyst, k_i is the interface partition coefficient, D is the solute diffusivity, ν is the kinematic viscosity, γ is the solid-to-liquid interfacial energy, ΔS is the entropy of fusion, γ_i^f is the fault energy, N_i^f is the number of faults of type i, μ is a parameter related to interface attachment kinetics, K is the thermal conductivity, and α is the heat diffusivity.

atively impossible to predict the behavior of one system on the basis of the performance of another since the variation of any one of the parameters or variables leads to large variability in the morphology of the growing crystals and thus in the resulting structure of the solid. Using the approach of Eq. (2.2), the problem becomes manageable and one can partition the total excess free energy driving the total reaction at any time into the partial excess free energies consumed by the various elemental f_i in the system as a function of time.[1] It is only by using the approach of Eq. (2.2) rather than Eq. (2.1) that this basic metallurgical problem has become manageable, i.e., we have reduced it to the simultaneous solution of 9 interrelated physics problems.

It should be made clear at this point that I am not proposing that the complete and successful description of some multiple-associated event in science or technology be derived completely by pencil and paper procedures. Nor am I implying a weakness of the basic philosophy of analytic procedures by raising this point. Rather, I am trying to be realistic about a basic characteristic of man that makes him interested in events that challenge his imagination and stretch his limitations. Stated another way, man generally seeks a successful system solution with knowledge of only a fraction of the pertinent information. If he knows all the facts needed for such a systems analysis, he is generally no longer vitally interested in the solution. Because of this condition, the total path to gaining a successful solution to such systems problems consists of two segments. The first segment is a scientific trajectory based upon the available information and the second segment is an empirical study of the system. As we shall see below, both segments are of vital importance to the success of the endeavor.

Figure 2 illustrates the ratio O.R./I.S. for the system in some hyperspace, where the coordinate axes are the independent variables $\varepsilon_1, \ldots, \varepsilon_j$, $X_1, \ldots, X_k$, and, for illustrative purposes, uses a three-dimensional figure where O.R./I.S. forms some surface with ε_i and X_i. Let our starting position for some state of the system be at point A and let one of the possible "success" locations be centered at B (others not shown on the diagram). The question we now ask ourselves is "Starting at A, how do we find B?" To answer this, we note that our scientific trajectory takes us to a point C which is located somewhere in a volume increment ΔV of this hyperspace which includes B. The size of ΔV depends upon the accuracy of our scientific analysis and the reliability of the input data. With point C as center, an empirical study is made of the surrounding volume V of the hyperspace that just includes B. Let us call this the "domain of credibility" for finding the success point with a reasonable expenditure of time and effort. If we

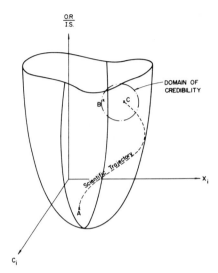

Fig. 2. Three-dimensional representation of O.R./I.S. correlation as a function of two of the natural system variables illustrating the complete path needed to proceed to the "success" location B from the starting state A.

think of this volume V as a hypercube of side d and wish to make experimental investigations at some average grid spacing λ, the number of experiments Q needed to map the behavior of this hypercube is given by

$$Q = \left(\frac{d}{\lambda}\right)^{p} \qquad (2.3)$$

where p is the total number of parameters and variables chosen as coordinates of this hyperspace (we have conveniently neglected the uncertainty of C and the success volume surrounding B). Taking the average time per experiment as τ_0 man-days, the total time τ for the empirical study is given by

$$\tau = \tau_0 Q \qquad (2.4)$$

The cost of the empirical study will be proportional to τ (to a first approximation).

To illustrate the importance of Eq. (2.4) suppose we are considering an elementary event where $p \sim 3$ and choose $\tau_0 \sim 1$, i.e., we are considering a typical physics problem. Let us suppose that we can afford only three man-years for the study, i.e., a typical Ph.D. thesis duration. Then from Eqs. (2.3) and (2.4) we find that $d/\lambda \sim 10$ and thus can develop and test fairly detailed models. For this example a fairly reliable pattern of behavior can be mapped and real scientific understanding generated. As p increases to about 10, as we may find in a typical engineering problem, we find that d/λ decreases to about 2 for the same τ/τ_0 so that only a very coarse grid of experimental points has been staked out in our hypercube and only the crudest of models can be tested with this amount of data.

Such a limited study at best is only able to rationalize behavior; it does not fulfill the requirements imposed by our definition of science (by using dimensionless groupings of variables, p can be significantly reduced). As p increases to about 20–30, such as one finds in many real events encountered in metallurgy, d/λ is only slightly larger than unity and only a narrow band of behavior is charted in our hypercube with this amount of data. If, by chance, the success point B is intersected by this study, only a single recipe results as the path or trail through the unknown surroundings. *We should begin to see one of the great advantages of computer simulation, i.e., the great reduction that is possible in the magnitude of τ_0.*

At this point we can begin to see why metallurgy has been largely an art up to the present day whereas physics has been a science for a few hundred years. We also can begin to ask, if it has taken this long to begin to transform metallurgy into a science how long will it be before the fields of medicine and sociology (for example) can be classified as sciences with the same level of reliability as present-day physics. Both fields are characterized by multiparameter, multivariable interaction events with large p and where, for medicine, the parameters are each a function of the body condition. In addition, for sociology the parameters are both a function of the environment and the evolution history of the environment (long time-constant phenomena). One might guess that several hundred years will be required for medicine and perhaps thousands of years for sociology if the Eq. (2.1) philosophy is followed for these systems of thought and practice. *However, the extensive use of simulation techniques on large-capacity, fast computers can greatly reduce this time.*

One might also begin to understand why metallurgists make good managers, businessmen, and politicians. It is because they have either intuitively or consciously learned to accept and handle the multiparameter, multivariable interaction problem. They tend naturally to seek a balance between a seemingly large number of forces or options that gives the best compromise of properties to the material or process (system). In this respect, their closest counterparts are the chemical engineers.

Before leaving this section, it is important to emphasize that *both* the scientific trajectory segment and the empirical segment are equally necessary parts of the path to understanding and achievement in these systems areas of endeavor. As the complexity of the system increases (p increases), the accuracy of the scientific trajectory must increase because the size of the "domain of credibility" is decreasing, i.e., the size that can be surveyed by a given number of man-years of effort.

Many people have hypothesized that, as our scientific knowledge increases, the delay time between scientific discovery and technological innovation should decrease. This is certainly true if p is held constant. However, as time passes and our scientific understanding grows, so also does the demand to control, within narrower limits, more variables and parameters in the systems event. Thus p tends to increase also so that the time (and cost) required from discovery to innovation does not generally decrease as our scientific knowledge increases. In fact, the time may significantly increase.

III. THE STUDENT AND HIS RESEARCH

At the moment an insufficient appreciation seems to exist, on the part of most students and some faculty, for the total system from specific phenomenon-oriented understanding to the ultimate application of knowledge. Thus perspective of and attunement to the larger picture is needed and an intellectual awareness of the important subroutines generated. Just as one knows that the failure of only one component in a space probe is often sufficient to abort the entire mission, it is necessary to instill in the student the wisdom of knowing that reliable prediction for the system requires reliable prediction for every single one of the subroutines and their interactions.

In the distant past, students treated their problem areas in the manner of Fig. 1a. More recently, the discrimination process of Fig. 1b became operative and students began to dig into the particular subroutines of the system with "laser-like" concentration to the exclusion of the total system. This has been a very natural psychological procedure, and it is important that a particular student develop detailed expertise in one of the subroutines of the system. However, the exclusiveness of this penetration or specialization threatens to decouple the system and they are in some danger of losing their prime function, i.e., of competently handling systems problems. It is time to progress to the next state of this evolutionary process exemplified by Fig. 1b, i.e., to maintain expertise in one of the subsets of the system, but to understand, work with, and appreciate the other subsets of the system, their interaction, and the overall function of the system.

This lets them know the ultimate value and relevance of their work to the immediate as well as to the long-range goals.

At the moment an inadequate mathematical awareness and mathematical skill exists among the student population in the materials science area. On the average they do not use mathematics as a comfortable and well-trusted tool to enlighten their understanding. This is a pity because we can recognize three distinct paths to knowledge: (1) total experimental, (2) theoretical and analytical based on idealized models and variable parameters, and (3) theoretical and numerical based upon much more exact models with specific parameters. As pointed out earlier, one needs to deduce the locations of certain "domains of credibility" wherein states of success lie; for this we need the second and third paths. To gain the final success point, we need to use the first path. Although the interactions between the subsets of our system are often of sufficient complexity that analytic solutions must relate to a too idealized model to be exactly relevant, the differential equations that govern the time–space change of our variables are well specified. Further, the boundary conditions are generally also well specified over the contours of the domain that contains our problem; thus a computer solution is definitely possible and will, I feel, become the most tractable path to the "domain of credibility" once the students have learned to use it creatively.

Because of a present lack of awareness relating to the uniqueness of this technical area, insufficient appreciation exists for the application of knowledge. We must begin to ask ourselves why many American companies are seeking B.S. and M.S. students rather than Ph.D. students. They say that it is because more use can be made of them. This means that our total education is out of balance—the more education you receive past some point, the less valuable you become. I'm sure that the fault lies with both the student and the company as already discussed. In the minds of many students, this function appears to have lost intellectual stature. Is this because we are breeding a sterile company of students or is it because they have been taught great skills without having been taught the perception of how to use these skills with assurance and pride to solve practical problems? I prefer to think that it is the latter and that the systems viewpoint on the part of both student and management will allow a practical problem to be phrased in such a way that a student can continue to grow as he generates the solution. I fear that the present manner of problem solving in some industries diminishes the student rather than builds him, and he tends to avoid this procedure like the plague. In essence, we have educated him too much to have him go back and solve real problems the "old" way, via

Fig. 1a; rather, we have to complete his education so that he can solve these real problems the "new" way, via Fig. 1b.

It is important that we ask, "What is the function of the student's research work?" It appears to have a threefold function: *The primary function is that it serve as a vehicle for unfolding the individual.* Every student has a special greatness in him—the problem is to help him unfold in a way that he can see it, he can begin to trust in it, and he can begin to build on it. In this way he grows in understanding and control of himself and his abilities. Second, the research must provide specific knowledge as a base for future work. This is often the entry into a particular employment area. Finally, it must serve to build confidence and provide some of the tools needed to solve real-life problems (professional and personal). The research problem serves as the focal point around which a one-to-one relationship between the professor and student exists. It is through this special three-body interaction that the unfolding process is stimulated.

To achieve the desired goal, the individual thesis research problems must be small enough for the student's present talents and tools to almost circumscribe (this generally means that it is a subroutine of a larger problem). It also must be significant enough to be worthy of his attention and that of his peers. This is the "muscle-building" that produces confidence and growth and the ability to become a problem-solver. It seems axiomatic that the act of doing a thing successfully augments both the ability to do that type of thing more easily and the appetite to do it more often.

The general departmental course work and research work must be designed in such a way that the student will not only be in tune with the systems frame of reference for perceiving his environment, but he will have the mathematical capabilities to analyze systems events and will present to society a spectrum of capabilities such as indicated in Fig. 3. He will exhibit a good background intensity I of ability over a very wide spectral range ν of scientific content (plus nontechnical content); this is his "generalist" component. Further, he will exhibit at least one extremely high spectral peak at some specific frequency ν^* relating to the position of one of the activities denoted in Fig. 1b; this is his "specialist" component. The specialist component gives him the "expert" identity and allows him to be uniquely useful and uniquely visible, which thus invites problems to come to him and give him the opportunity of growing further by solving them. The generalist component allows him to couple with other specialists in his environment in a meaningful and cooperative way to both pump additional insight into his ν^*-mode and to effectively treat large systems event problems. One great dilemma that faces us is that the course offerings

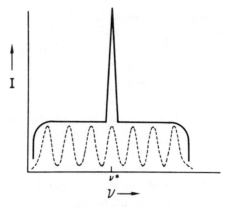

Fig. 3. Schematic representation of output intensity I of students as a function of an "effective" frequency of technical (and non-technical) competence.

in most departments are isolated bands of knowledge that tend not to overlap each other (as indicated by the dashed curves in Fig. 3). It is necessary that we generate a means of synthesizing this information to yield the continuous broadband awareness indicated in Fig. 3.

One path that promises to be very useful in this regard is the development of computer simulation courses all students would take. Systems event problems would be programmed for the computer, the student needing to put all the relevant physics into the various subelements of the system and couple them in the appropriate fashion. By varying certain parameters in various subroutines, the output spectrum of the total system can be studied. In this fashion vicarious experience can be gained concerning the "systems event." The student would learn not only the techniques of dealing with this kind of problem but, more important, he would realize the subtle interplay of forces in real problems and should find the confidence to meaningfully face and resolve this class of problems in his personal and professional life.

IV. EXAMPLES OF SYSTEMS EVENTS IN THE CRYSTALLIZATION AREA

A. An Overview of the Scientific Subroutines

To illustrate what fields of science and what general knowledge must be brought to bear upon a reliable analysis of a crystal growth problem, let us take the example mentioned earlier and say that we wish to control the structure of a solid by isothermally freezing a small volume of liquid alloy. By structure we shall mean grain size and shape, degree of microscopic and macroscopic chemical segregation, etc.

For this "overview" we will be satisfied with a phenomenological description of the important processes in terms of "lumped" parameters. The conventional macroscopic variables that we either set or control are (1) bath composition C_∞, (2) bath cooling rate $\dot{T}$, and (3) the shape of the container holding the liquid. Let us proceed with the process description by stages.

1. As the liquid is being cooled, we need to know the magnitude of the driving force for solid formation ΔG at any bath temperature T. This can be expressed as

$$\Delta G = f_1[\Delta H, T_L(C_\infty)] \qquad (4.1)$$

where f_1 refers to the appropriate functional relationship between the latent heat of fusion ΔH and the liquidus temperature $T_L(C_\infty)$. Thus we see that phase equilibria data is one prerequisite. The material parameters needed for this area of study are listed and defined in Table I.

2. As the bath undercooling ΔT increases with time t we need to know the undercooling at which particles of solid begin to form and also their density. Thus we must evaluate the nucleation frequency I, which can be most simply expressed as

$$I = f_2(N_0, \Delta T_C, \dot{T}, t, C_\infty) \qquad (4.2)$$

where f_2 represents the appropriate functional relationship, N_0 is the number of atoms in contact with the foreign substrate that catalyzes the nucleation event, and ΔT_C is a parameter that defines the potency of the catalyst.

3. When the crystals illustrated in Fig. 4 begin to grow at some velocity V, solute partitioning will occur at the interface since the equilibrium concentration of solute in the solid C_S is different than the concentration in the liquid at the interface. Thus the concentration of solute in the liquid at the interface C_i must be determined and can be represented by a functional relationship of the form

$$\frac{C_i}{C_\infty} = f_3(V, k_i, D, \delta_C, S, t) \qquad (4.3)$$

where k_i refers to an interface solute partition coefficient that is generally different from k_0, D is the solute diffusion coefficient, δ_C refers to the

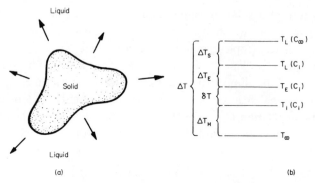

(a)   (b)

Fig. 4. Left: Illustration of a crystal growing from a super-cooled liquid. Right: The important temperatures in a growth process. The magnitudes of the temperature differences indicate the degree of solute diffusion, capillarity, kinetic, or heat transport control.

solute boundary layer thickness at the crystal surface, and S refers to the shape of the crystal.

4. In order to evaluate δ_C in Eq. (4.3), it is necessary to consider the hydrodynamics of the fluid. The fluid will generally exist in some state of motion whether the driving force is applied by external means or arises naturally due to density variations in the fluid. We can consider the fluid far from the crystal–liquid interface to be moving with some relative stream velocity u_∞ due to the average fluid body forces. The fluid motion will aid in the matter transport of solute away from the crystal into the bulk liquid and cause a lowering of C_i. We find that δ_C can be expressed as

$$\delta_C = f_4(V, \nu, D, S, u_\infty, t) \qquad (4.4)$$

where ν is the kinematic viscosity of the fluid.

The portion of the total undercooling consumed in driving the solute transport ΔT_S is given by

$$\Delta T_S = T_L(C_\infty) - T_L(C_i) \qquad (4.5)$$

5. Because the growing crystal is small in size, has curved surfaces, and often contains nonequilibrium defects, the solid contains a higher free energy than the solid considered in generating a phase diagram that we use as our standard state in the overall treatment. Thus the equilibrium melting temperature for such a solid is lowered by an amount ΔT_E compared to that for the equilibrium solid. We find that the portion of the total under-

cooling consumed in the production of nonequilibrium solid $\varDelta T_E$ can be expressed as

$$\varDelta T_E = T_L(C_i) - T_E(C_i) = f_5(\gamma, \varDelta S, S, \Sigma_i \gamma_i{}^f, \Sigma_i N_i{}^f) \qquad (4.6)$$

where $T_E(C_i)$ is the equilibrium interface temperature for interface liquid concentration C_i, γ is the solid–liquid interfacial energy, $\varDelta S$ is the entropy of fusion, $\gamma_i{}^f$ is the fault energy for defects of type i, and $N_i{}^f$ is the number of type i.

6. Next, because the crystal is growing, a departure from the equilibrium temperature $\varDelta T_K$ must exist at the interface in order to produce a net driving force for molecular attachment to the growing solid. At sufficiently large departures from equilibrium, the molecules can attach at any interface site and lower the free energy of the system. However, at small departures from equilibrium, molecular attachment at random interface sites generally leads to an increase in the free energy of the system and thus such interface attachment will not occur as a spontaneous process. Rather, in such an instance molecules become a part of the solid only by attachment at layer edge sites on the interface and one must consider the various mechanisms of layer generation on the crystal surface. The portion of the total undercooling consumed in driving this interface process $\varDelta T_K$ can be expressed as

$$\varDelta T_K = T_E(C_i) - T_i = f_6(V, \mu_1, \mu_2, S, t) \qquad (4.7)$$

where T_i is the actual interface temperature and where μ_1 and μ_2 are lumped parameters needed to specify the interface attachment kinetics for the various attachment mechanisms.

7. Finally, since the crystal is growing it must be evolving latent heat and the interface temperature T_i must be sufficiently far above the bath temperature T_∞ to provide the potential for heat dissipation to the bath. That portion of the total undercooling consumed in driving the heat dissipation $\varDelta T_H$ can be expressed as

$$\varDelta T_H = T_i - T_\infty = f_7(K, \alpha, \varDelta H, V, S, t) \qquad (4.8)$$

where K refers to the thermal conductivity and α refers to the thermal diffusivity.

The foregoing has been a description of the subdivision of the total bath undercooling $\varDelta T$ into its four component parts, i.e.,

$$\varDelta T = \varDelta T_S + \varDelta T_E + \varDelta T_K + \varDelta T_H \qquad (4.9)$$

Equation (4.9) is called the "coupling equation" and illustrates the fact that these four basic elements of physics enter every crystal growth situation and are intimately coupled through this constraint. However, for different materials certain of the components of Eq. (4.9) tend to dominate the phase transformation. In the growth of metal crystals from a relatively pure melt, $\Delta T_H \sim \Delta T$ so that this case is largely controlled by heat flow. During the growth of an oxide crystal from a melt of steel, e.g., $\Delta T_S \sim \Delta T$ so that the growth is largely diffusion controlled. During the growth of a polymer crystal from a well-fractionated polymeric melt, $\Delta T_K \sim \Delta T$ and the growth is largely controlled by the kinetics of interface attachment. Finally, during the growth of a lamellar eutectic crystal, $\Delta T_E \sim \Delta T/2$ and the growth is to a large degree controlled by the excess free energy of the solid (due to the formation of α/β phase boundaries). By considering Fig. 5, which is a plot of crystal growth velocity as a function of time, we find that at small times ΔT_E in Eq. (4.9) dominates the crystal's growth and thus plays an overriding role in its morphology. At large times ΔT_S and ΔT_H in Eq. (4.9) dominate the crystal's growth and lead to very different morphologies. At intermediate times all four factors play significant roles in the shape adopted by the crystal.

There is little doubt that crystal morphology plays a significant role in the resultant crystal perfection and that this morphology is largely determined by the subtle interplay of the factors already discussed. However, the prediction of crystal shape with time is a problem that we have been unable to solve in any general way. This arises because the problem thus

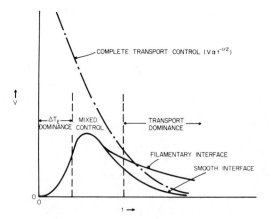

Fig. 5. Schematic of particle growth velocity V versus time t illustrating the regions where various mechanisms are dominant.

far stated is, in general, not completely specified. Knowing V and S, C_i and T_i can be completely determined in terms of C_∞ and T_∞, respectively, and also ΔT_S, ΔT_E, ΔT_K, and ΔT_H are completely determined in the general mathematical sense. However, we do not know either V or S. If $S(t)$ is specified, Eq. (4.9) can be used to determine $V(t)$. We are in need of an additional condition to completely specify the problem and thus provide simultaneous prediction of both $V(S)$ and $S(t)$.

We shall find that our extra condition is generated by considering the response of the growing crystal to shape perturbations. It can be easily shown that the various elements of the interface will always be subjected to fluctuations in ΔT and in its component ΔT_i's. Thus, given sufficient time we should always expect that shape distortions will have a finite probability of forming on the crystal surface and that the crystal will evolve to that shape which is most stable in the environment that allows such perturbations. With the addition of our perturbation response equation in the general form

$$V^*(S) = f_8(S^*, \Delta T_S, \Delta T_E, \Delta T_K, \Delta T_H) \qquad (4.10)$$

the most stable crystal shape S^* and the corresponding growth velocity at various points on the crystal surface V^* may be evaluated from Eqs. (4.9) and (4.10).

At this point in the problem we can, in principle, specify the solute, temperature, and fluid velocity fields throughout the system. We thus have the potential to specify the chemical inhomogeneity of the final solid and the types of compound formation that should occur in the highly solute-enriched liquid that is last to freeze. The stress distribution due to both constitutional and thermal variations in the as-solidified alloy should also be specifiable, and thus the defect generation in the solid as it cools to a particular temperature may be predicted.

B. Technological Understanding of Ingot Defects

From our technological experience of the past few years, it has become apparent that four primary casting defects are limiting technological success in the foundry area. These can be categorized under the headings (1) micro- and macroporosity, (2) micro- and macrosegregation, (3) inclusions, and (4) hot tears. Careful consideration has been given recently to the possible causes of these defects using the best insights available from the fundamental understanding that exists in the solidification area, and we are able to make educated guesses as to the most probable causes of the defects in castings. This enhanced understanding indicates that a sensible path exists for the development of complex alloy castings without the abovementioned defects being present to any serious degree.

1. The driving force for microporosity comes from the volume change of the alloy on freezing, i.e., the solid is generally more dense than the fluid and a certain amount of shrinkage must occur. The location of the microporosity is in the interdendritic and grain boundary regions, and the voids form because an insufficient head of pressure exists to feed liquid through these channels at the required rate during the freezing of the casting. The longer the feeding channels, the more gassy the fluid, the more viscous the fluid, the greater the degree of microporosity. This means that, for fixed size and shape of casting, the greater the viscosity of the liquid, the greater the thermal conductivity of the solid, the greater the freezing range of the alloy, the greater the degree of microporosity in the casting.

The driving force for macroporosity also comes from the volume change on freezing and is augmented by any propensity for gases to come out of solution at reduced pressures. The location of the macroporosity is in regions of the casting where the formation of solid was nonuniform so that isolated pockets of liquid had become surrounded by grains that were growing dendritically. The demand of these interdendritic regions for the flow of fluid from such pockets plus the inability of the main body of fluid to, in turn, feed the isolated pockets combine to create a void of macroscopic size in such regions. The incidence of such events is directly traceable to this nonuniform formation of solid. It can be qualitatively shown that the probability of forming such critical pockets of fluid increases as the nucleation capabilities of the alloy decrease and increases with those other factors that increase the microporosity.

2. The driving force for microsegregation is the partitioning of solute between solid and liquid during freezing because the solubility of solute in the solid is different than in the liquid. The smaller this solid solubility, the smaller the solute distribution coefficient k_0 and thus the greater the degree of chemical microsegregation. Macrosegregation occurs via the same driving force but requires, in addition, that macroscopic fluid motion occurs so that solute which is partitioned at the interface is not retained in a thin boundary layer at the surface but is mixed with the bulk liquid. When freezing occurs in an orderly fashion progressing from one location to another some distance away at some later time, such fluid mixing causes the last part to freeze to have a different solute content than the first part of the fluid to freeze—this is macrosegregation.

3. What we might call poor housekeeping practices will lead to the formation of relatively large inclusions in castings. However, an important source of inclusions is the chemical reaction between alloying constituents in the melt during the freezing of the casting. As solid is formed, the various solute elements are partitioned into the remaining liquid. As solute enrichment occurs, the activity of the reacting constituents increases and the inclusion formation increases. The frequency of inclusion formation increases with the partitioning potential of the alloying and minor constituents and tends to increase as the freezing range of the alloy increases.

4. The driving force for hot tearing comes from nonuniform thermal contraction of the casting as it freezes and is cooled, i.e., the outside wants to contract because it is cold whereas the inside is still hot and does not need to contract. This puts the outer regions of the casting under tensile stress, which causes the shell of solid to tear if the stresses exceed the strength of this solid plus liquid mixture at the particular temperature. The tear will be rehealed if fluid feeding from the main bath to the torn region can readily occur. The magnitude of the stresses increases with the steepness of the temperature gradient in the solid shell, i.e., increases with rate of cooling and sharpness of a corner in a casting. Both the strength of the solid shell and the probability of tear rehealing will decrease as the freezing range of the alloy increases. Thus the frequency of hot tear defects will increase as the rate of chilling increases and as the freezing range of the alloy increases.

Below, the outlines of a systems analysis of the hot tearing defect is presented. Following this procedure the student could readily identify the boxes in Fig. 1 plus the parameters and variables that are involved in understanding and controlling the other three defects.

The integrity I of a casting can be considered to be a function of the stress applied to the casting σ_A and the failure stress of the casting σ_f, i.e.,

$$I = f_1[\sigma_A(t), \sigma_f(t)] \tag{4.11}$$

where t refers to time both during the formation of the casting and later in application. During casting the applied stress is due to thermal contraction and may be given by

$$\sigma_A = f_2(\dot{T}, S, \beta, E, T - T_L) \tag{4.12}$$

where $\dot{T}$ is the cooling rate, S is the shape of the casting, β is the thermal expansion coefficient, E is the bulk modulus of the solid, and T is the actual temperature in the casting.

During casting the failure stress depends very critically upon the morphology $M_S{}^L$ of the liquid–solid mixture, the volume fraction g of primary solid, the volume fraction v^q of any additional phase present (voids, inclusions, etc.), and the intrinsic fracture strength σ_F of the aggregate mixture, i.e.,

$$\sigma_f = f_3(g, M_S{}^L, \Sigma_q v^q, \sigma_F) \tag{4.13}$$

Even after casting without a failure, the amount and the distribution of the v^q may make the casting useless for its intended application. In turn, the v^q that form during freezing depend upon $M_S{}^L$, the concentration of the freezing interface $C_L{}^I$, the temperature of the freezing interface T_L, the free energy of formation ΔF^q of phase q, and the nucleation probability $P_N{}^q$ of forming phase q, i.e.,

$$v^q = f_4(M_S{}^L, C_L{}^I, T_L, \Delta F^q, P_N{}^q) \tag{4.14}$$

One next finds that $C_L{}^I$ and T_L are connected through the Gibb's equilibria equations. They are both dependent upon the fraction of solid g and the thermodynamic activity coefficients in the solid $\gamma_S{}^j$ and in the liquid $\gamma_L{}^j$, i.e.,

$$C_L{}^I = f_5(T_L, \Sigma_j \gamma_S{}^j, \Sigma_j \gamma_L{}^j, g, M_S{}^L) \tag{4.15}$$

$$T_L = f_6(\Sigma_j C_L{}^j, \Sigma_j \gamma_S{}^j, \Sigma_j \gamma_L{}^j) \tag{4.16}$$

Finally, ΔF^q depends upon the concentration and temperature of the liquid, the activity coefficients, and the standard free energy of formation of q, $\Delta F_0{}^q$, i.e.,

$$\Delta F^q = f_7(T_L, \Sigma_j C_L{}^j, \Sigma_j \gamma_L{}^j, \Delta F_0{}^q) \tag{4.17}$$

With the foregoing procedure we have blocked out or modeled the system to a level that, although not perfectly exact, is quite satisfactory for obtaining a good level of understanding of the problem and a good degree of reliability with respect to predicting the domain of credibility for finding sound castings free from hot tear defects.

C. Solute Distribution in Pulled Crystals

In crystals prepared by the Czochralski technique (crystal pulling from a crucible), the solute concentration can be completely uniform in both the radial and the longitudinal direction or it can contain any or all of the following irregularities: (1) longitudinal gradients, (2) radial gradients,

(3) bands of excess solute parallel to the interface, or (4) channels of solute in a cellular pattern perpendicular to the interface. The solute concentration in the solid is found to be a function of the freezing velocity V, time t, bulk liquid concentration C_∞, phase diagram distribution coefficient k_0, solute boundary layer thickness δ, thermal fluctuations in the melt δT, and the shape of the interface S, i.e.,

$$C_S = f_1(S, \delta, k_0, V, t, \delta T, C_\infty) \tag{4.18}$$

The values of δ and S are interconnected via the convective flow and the heat transport. In addition the crystallography of the interface has a strong influence on its shape. The natural convection due to the radial temperature gradient in the system draws the fluid up the sides of the vertical crucible along the top and down the center. The velocity of this natural flow depends upon the temperature difference between the top and bottom of the crucible $T_h - T_c$, the dimensions of the crucible R and L, the kinematic viscosity v, the thermal expansion of the fluid β, the radius of the crystal r, and the shape of the interface S. Except for the last two terms, these variables can be combined into the Grashof number G_r. The forced convection, due to the rotation of the crystal at a rate ω, draws fluid up the center of the crucible outward along the top of the fluid and down the outer walls. A conflict tends to occur between the flow patterns due to forced and natural convection wherein natural convection dominates at small values of ω with forced convection dominating at large values of ω. At in-between values of ω, one often finds eddies and quiet domains of fluid, which is not the best condition for homogeneous mixing. Thus we find

$$\delta = f_2(G_r, \omega, r, S, V, D) \tag{4.19}$$

The shape of the interface depends on the thermal properties of the liquid and solid K^j and α^j plus the latent heat ΔH as well as crystallography X. In addition, it depends upon δ and whether natural convection dominates (S is concave to solid) or the then forced convection dominates (S is concave to liquid), i.e.,

$$S = f_3(G_r, \omega, \delta, X, V, \Delta H, K^j, \alpha^j) \tag{4.20}$$

The banding of the solute is caused by a fluctuation of the freezing velocity ΔV that comes about, most generally, from temperature oscillations in the melt. These oscillations are caused by inverse density gradients (low density on bottom, high on top) driving the fluid into cellular motion.

The greater the temperature difference from top to bottom of the cell δT_c, the greater the oscillation of temperature δT transported to the interface. The smaller the thermal conductivity and diffusivity of heat in the liquid, the greater δT. Finally, the presence of a strong dc magnetic field H can diminish the convection, as can a centrifugal motion of the fluid at frequency W, and thus δT in conducting liquids, i.e.,

$$\delta T = f_4(T_h - T_c, \delta, K^j, \alpha^j, \nu, H, W) \tag{4.21}$$

Finally, the segregation of solute into cell boundaries perpendicular to the interface depends upon the presence of constitutional supercooling (CSC) in the liquid, the thermal properties, the interfacial energy γ, and the molecular attachment kinetics.

Chapter 2

Lectures on Large-Scale Finite Difference Computation of Incompressible Fluid Flows

Jacob E. Fromm

IBM Research Laboratory
San Jose, California

I. THE DIFFERENTIAL EQUATIONS

A. Introduction

In the present section we shall attempt to describe, through example, the essentials of numerical computation of time-dependent, nonlinear fluid flows. The case in consideration will be that of incompressible flow with viscosity, described in terms of a vorticity and streamfunction. The discussions have been simplified so that the overall view of the computation procedures can be emphasized. Refinements of the individual areas, or subprograms, are presented in succeeding sections where the methods are brought up to date. Along with recommended reading the included material should permit the construction of a working program. Alternatively, the outline provided should lend itself to expansion into other areas of numerical computation of initial-boundary value problems.

It is not our objective here to lay a foundation in classical hydrodynamics since this would lead us far from the pertinent points of numerical computation, particularly as related to finite difference computation. It should be understood that finite difference methods for nonlinear equations are generally regarded as a very direct approach to solution of initial-boundary value problems; it therefore behooves us to take an experimentalist's viewpoint. We look upon the partial differential equations as statements of conservation laws and seek an insight into the role played by each term in development of flow behavior. Rigorous interpretation of terms that should be deleted or included relates to the application in question as do the variables used in describing the problem.

We must of necessity speak of situations where we have a continuum behavior so that functional values and their derivatives can be evaluated without ambiguity. This leads to conflict with an often used analytic approach to problems, namely, that of permitting mathematical discontinuities. Typical of the contrast in viewpoints are compressible flow computations involving shock waves. The classical approach is to regard the shock as a discontinuity. Numerically it is assumed to be a sharp variation but continuous. In nature it *is* continuous if we look at it closely and, while it is often sharper than our numerical representation can resolve, our continuous description does permit us to conduct valuable numerical simulation experiments. It is the notion of simulation that further suggests the experimentalist's viewpoint. Because the exact entropy generating processes of shocks cannot and need not be included in gaining the knowledge we seek, we can empirically lump these unknown processes into adjustable terms that can provide the continuity necessary to computation. Even on the computer we must set aside our idealistic notions of knowing all there is to know about a given situation. We gain the advantage of an ability to treat nonlinearities but must accept the fact that the solutions are inherently inexact. We accept this inexactness in the same sense that the experimentalist accepts his measurements as inexact. It represents a revision of our thinking as theoreticians, i.e., we abandon the exact analytic solution of idealized equations for the approximate solution of more complete equations. Actually the beauty of the former is often superficial because of its limited value in the real world.

B. Numerical Solution of Laplace's Equation

As a first example of numerical solution we consider the solution of Laplace's equation

$$\nabla^2 \psi = 0 \tag{1.1}$$

Our book knowledge suggests consideration of a well-defined problem in which boundaries correspond to coordinate lines or surfaces of a system that is separable and thereby is amenable to well-known techniques of solution. The equation is elliptic and therefore requires boundary conditions on a closed boundary. The Laplacian being zero means that there are no extremals in the field, i.e., the maximums or minimums of the ψ distribution must occur on the boundary. One may use finite differences here simply to gain insight into the nature of solutions that satisfy (1.1). In two-dimensional rectangular coordinates we may write

$$\frac{\partial^2\psi}{\partial x^2} + \frac{\partial^2\psi}{\partial y^2} = \frac{\partial}{\partial x}\left(\frac{\partial\psi}{\partial x}\right) + \frac{\partial}{\partial y}\left(\frac{\partial\psi}{\partial y}\right)$$

$$= \lim_{\substack{\Delta x \to 0 \\ \Delta y \to 0}} \frac{(\psi_1 - \psi_0)/\Delta x - (\psi_0 - \psi_3)/\Delta x}{\Delta x} + \frac{(\psi_2 - \psi_0)/\Delta y - (\psi_0 - \psi_2)/\Delta y}{\Delta y}$$

where the numbers are related as in Fig. 1.

If we take $\Delta x = \Delta y$, then we may say

$$\psi_0 \approx \frac{\psi_1 + \psi_2 + \psi_3 + \psi_4}{4} \tag{1.2}$$

or ψ is everywhere the average of its nearest neighbors. Now the notion of the absence of extremals becomes clear as does the necessity for prescription of boundary conditions on a closed boundary. Every value depends on its neighbors, which in turn depend on their neighbors, etc., until the boundary is reached. If boundary values are given, it becomes evident that if we successively and repeatedly take averages at a discrete set of points as in Fig. 1, we can ultimately satisfy (1.2) at all points. Prescribed functional values on the boundary (Dirichlet conditions) are easy to satisfy in

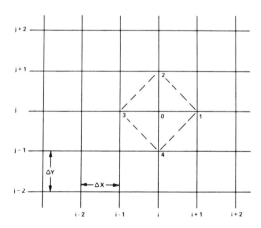

Fig. 1. Labeling of the computation grid.

this way, but intuitively we can see that a condition on the first derivative (Neumann conditions) will be somewhat more difficult to satisfy and will take more repetitions with (1.2). Also, Neumann conditions cannot be prescribed on all boundaries; we must fix at least one point for unique solution of the differential problem. Numerically a renormalizing procedure is used to keep the functional values within some bounds. Usually we are interested only in the derivatives of the function ψ and, hence, the values are used in a relative sense only.

C. The Inclusion of a Source Term and the Potential Solution

Our next step beyond Laplace's equation is to make (1.1) inhomogeneous by inclusion of a source term. Let us assume that Laplace's equation is satisfied for some circular region with fixed values (say $\psi = 0$) at boundary points. However, let ψ satisfy

$$\nabla^2 \psi = -\omega \tag{1.3}$$

at some single interior point. It is easy to visualize that the solution $\psi = 0$ everywhere is now modified to give a maximum in the region if ω is positive. This follows simply from our notions of the second derivative properties of the one-dimensional counterpart to (1.3). A point value of ω with Laplace's equation satisfied elsewhere has the physical significance of vorticity. Vorticity induces a circular motion (a vortex) centered at the point, and ψ represents the streamfunction.

Now we take

$$u = \frac{\partial \psi}{\partial y} \qquad v = -\frac{\partial \psi}{\partial x} \tag{1.4}$$

where u and v are velocity components in the x and y direction, respectively. In the example discussed above we visualize streamlines as contours about the center of rotation. The gradient (derivative taken normal to the contour lines) gives the speed, and the tangent to the contours gives the vector velocity direction.

The point vortex is called a "potential vortex" because indeed Laplace's equation is satisfied everywhere (excluding boundaries that may be at infinity) except at the point where ω is nonzero. Flows without vorticity are potential flows. We could, of course, regard the single point as a boundary point, and then indeed we would have a potential solution in the strict sense. We would come out with the same solution if we picked the appropriate value for ψ at the boundary point.

We can consider the inverse situation where we have a given boundary value problem for which Laplace's equation is satisfied at all interior points but we wish to find the distribution of vorticity on the boundaries that corresponds to the solution initially defined in terms of the stream function alone. This is a point pertinent to our later considerations of determining a time history of a vorticity distribution that will be, at the outset, defined only on rigid boundaries. Here we must bring in the notions of fluid behavior at a solid wall since the vorticity at the wall may be a function of the interfacing materials. This is completely analogous to stating that the charge distribution at the plate–dielectric interface of a condenser may depend upon the nature of the contacting surfaces. In engineering practice we take the boundary condition at rigid surfaces to be "no-slip," i.e., the tangential fluid velocity is zero. At a horizontal wall we would, of course, have v (the normal velocity) vanish, and thus ψ would be a constant (a streamline). But with no-slip the tangential velocity condition must be taken into account. For the potential problem this latter condition is supplementary since the solution would be completely determined by the condition on the normal velocity. Here, as in the shock problem, we have to deal with a velocity step function giving an infinite normal derivative for the tangential velocity. Thus

$$\frac{\partial^2 \psi}{\partial y^2} = -\omega \to \infty$$

or the vorticity is infinite. In finite differences the step function of velocity must be spread at least over one mesh length; thus we take a different route than that of analysis. This, incidentally, permits us to make a smooth transition from an initial solution into a time-varying solution in which viscous effects can come into play to remove the singular behavior at the wall. Physically we know the infinity condition does not arise, but to examine things on that scale would require a problem formulation that would deal with atomic particles. It is preferable not to argue a moot point such as this and to recall our limitation to discuss continuous behavior. Fortunately, finite difference expressions can be found that allow us to have consistency in the formulation of the boundary conditions with the only limitation being the coarseness of grid. We would expect the solutions in the limit of infinitesimal grids to approach the differential problem solution.

At this stage we know in a rough way how to find approximate potential solutions to any well-defined problem. There are countless techniques for obtaining rapidly convergent and highly accurate results. Often, however, the techniques are restricted to only certain types of boundary condition combinations, and their speed of computation may very likely be an inverse function of their generality. It often pays to go back to the fundamental form (1.2) as a backup for any sophisticated method devised and also as a starting point for developing intuitive notions about a given problem.

D. The Time-Dependent Vorticity Field Due to Diffusion and Convection

Our next interest is to see what happens to the charge on a condenser plate or the vorticity on a boundary surface. One can imagine a diffusing process taking place. We can write an equation for the time variation of ω as follows:

$$\frac{\partial \omega}{\partial t} = \nu\left(\frac{\partial^2 \omega}{\partial x^2} + \frac{\partial^2 \omega}{\partial y^2} \right) \tag{1.5}$$

We have already had some experience with the Laplacian and the averaging process that is inherent in it. In (1.5) it causes smoothing of the ω distribution or a spread of values away from the boundaries if we are visualizing a change from a prescribed boundary distribution. The factor ν governs the rate at which the averaging process takes place. Equation (1.5) is the same as the parabolic heat conduction equation, where ν is the heat conduction coefficient. If ω is the vorticity, the coefficient is the viscous diffusion stemming from molecular attractions. These molecular attractive forces cause various layers of fluid moving at different velocities to tend toward more equal velocities. The vorticity is thus a measure of the difference in velocity of the layers. If the fluid is a gas, the viscous effects stem from molecular collisions and momentum is transferred to slower moving layers. In a gas the viscosity increases with temperature because molecular agitation causes greater interaction of fluid layers. In liquids the molecular attraction forces are diminished by increases in temperature because higher energy levels correspond to more loosely bound states. We thus can visualize vorticity migrating from the initial distribution on the boundary by this diffusional process ultimately toward a state where the distribution is one of linear variation. When the variation becomes linear, the Laplacian goes to zero and the diffusion process ceases.

The diffusion process is not the only one whereby a transfer of vorticity can take place. The distribution can become modified by convection. Once vorticity exists in the fluid region, near the boundary, it can be carried by the fluid motions just as dust may be lifted from the surface by a gust of wind. In fact, if we ignore the diffusive process or assume it is very small except where gradients are very large (at the boundary) we find that the distribution is transferred by convection alone, just as in the case of dust particles. However, the vorticity is dynamically interacting in that its distribution both governs and is governed by the flow. Numerically, we alternately compute the distribution of vorticity based on the flow field and then find the new flow field corresponding to the updated vorticity distribution.

There are two ways of studying the convection of a variable distribution. Ignoring diffusional influences (conduction in the case of heat), we can assign our variable values to particles. In such a case

$$\frac{d\omega(x, y, t)}{dt} = 0 \tag{1.6}$$

and it is only necessary to keep track of the particle position as a function of its velocity thus:

$$x = x_0 + u\Delta t$$
$$y = y_0 + v\Delta t \tag{1.7}$$

This is called the "Lagrangian formulation of fluid motion." We could in this way reconstruct the distribution at various time increments from some initially known distribution if the velocities were somehow known. We could, e.g., consider a given distribution of ω (known everywhere) and then discretize this information to a fixed space and at time $t = 0$, letting the ω values be the average over a grid interval. Now assuming known velocities at the grid points, we could move these selected particles knowing that ω remains fixed for each particle. Finally, we could construct the distribution at $t = \Delta t$, again discretized in value but now with variable spacing. We could continue with this variable spacing if we always knew the particle velocities and could state to what interval the given functional values applied as an average. Some sort of mass average instead of space average would be more appropriate in such a description.

Suppose now that instead of continuing with a variable spacing of known values we perform an interpolation to reestablish the values on our original evenly spaced grid. We could do this at each time step and be using new particles for each consideration of (1.7). Note that our grid now becomes a fixed one through which the variable distribution is moving.

Thus (1.6) becomes

$$\frac{\partial \omega}{\partial t} + \frac{\partial \omega}{\partial x}\frac{\partial x}{\partial t} + \frac{\partial \omega}{\partial y}\frac{\partial y}{\partial t} = 0$$

or

$$\frac{\partial \omega}{\partial t} + u\frac{\partial \omega}{\partial x} + v\frac{\partial \omega}{\partial y} = 0$$

The partial derivative with respect to time is the change with time of ω with x and y fixed (a fixed point on our grid). Thus we recognize the added terms as the changes in ω because of flow past the grid point.

In the above we have outlined a means of developing a difference approximation method for the nonlinear convective terms. It is not the technique generally used to develop the schemes, but perhaps it should be. We can roughly judge the accuracy of the method by the order of our interpolation formula. We can use any interpolation method, but problems of numerical stability must be taken into account. One does not usually discuss an interpolation scheme in terms of numerical stability, but this is a valid and meaningful point of view to take.

E. The Dynamic Equations and Scaling

We will conclude this section by summarizing and then scaling our equations. The vorticity in the preceding formulation was influenced by convection and diffusion. We may write the equation for it as

$$\frac{\partial \omega}{\partial t} + u\frac{\partial \omega}{\partial x} + v\frac{\partial \omega}{\partial y} = \nu\left(\frac{\partial^2 \omega}{\partial x^2} + \frac{\partial^2 \omega}{\partial y^2}\right) \qquad (1.8)$$

Further, the vorticity acted as a source term for evaluating a stream function through a Poisson's equation; thus

$$\frac{\partial^2 \psi}{\partial x^2} + \frac{\partial^2 \psi}{\partial y^2} = -\omega \qquad (1.9)$$

and finally it was indicated by the definition of the stream function that

$$u = \frac{\partial \psi}{\partial y} \qquad v = -\frac{\partial \psi}{\partial x} \qquad (1.10)$$

so that (1.8), (1.9), and (1.10) represent a closed set of four equations for four unknowns.

Nondimensionalizing the equations is the first step in establishing scaling parameters. Let the space coordinates be relative to a distance d and the velocities be relative to a known velocity u_0. The time scale becomes d/u_0. We can write

$$t' = \frac{u_0 t}{d}$$

That is, the new unit of time (t') is the time it takes to travel the reference distance at the reference velocity. Our new velocities (u') are in units of u_0, and our new distances (x', y') in units of d. Since ω has units of velocity divided by distance,

$$\omega = \frac{u_0}{d}\,\omega'$$

The stream function has units of velocity times distance, hence

$$\psi = du_0\psi'$$

Thus the form of (1.9) and (1.10) does not change in terms of the dimensionless variables. Now

$$\frac{\partial\omega}{\partial t} = \left(\frac{u_0}{d}\right)^2 \frac{\partial\omega'}{\partial t'} \qquad u\frac{\partial\omega}{\partial x} = \left(\frac{u_0}{d}\right)^2 u'\frac{\partial\omega'}{\partial x'}$$

and

$$\frac{\partial^2\omega}{\partial x^2} = \left(\frac{u_0}{d}\right)\frac{1}{d^2}\frac{\partial^2\omega'}{\partial x'^2}$$

Thus

$$\frac{u_0}{d}\left(\frac{\partial\omega'}{\partial t'} + u'\frac{\partial\omega'}{\partial x'} + v'\frac{\partial\omega'}{\partial y'}\right) = \frac{\nu}{d^2}\left(\frac{\partial^2\omega'}{\partial x'^2} + \frac{\partial^2\omega'}{\partial y'^2}\right)$$

Dropping primes and defining

$$R = \frac{u_0 d}{\nu}$$

we have

$$\frac{\partial\omega}{\partial t} + u\frac{\partial\omega}{\partial x} + v\frac{\partial\omega}{\partial y} = \frac{1}{R}\left(\frac{\partial^2\omega}{\partial x^2} + \frac{\partial^2\omega}{\partial y^2}\right)$$

R is called the "Reynolds number" and is probably the most well-known scaling parameter in fluid flow calculations. Scaling allows us to study a variety of similar physical situations without specifying the units of measurement.

F. Suggested Reading

1. P. M. Morse and H. Feshbach, *Methods of Theoretical Physics*, McGraw-Hill Book Co., Inc., New York, 1953, Volume I, Chapter 6, Sections 1 and 2, 676–706.
2. B. Alder, S. Fernbach, and M. Rotenberg, *Methods in Computational Physics*, Academic Press, New York, 1964, Volume III, Chapter 10, 346–382.
3. R. D. Richtmyer, *Difference Methods for Initial-Value Problems*, Interscience Publishers, Inc., New York, 1957, Chapter 10.
4. L. Collatz, *The Numerical Treatment of Differential Equations*, Springer–Verlag, Berlin, 1960, Chapters 4 and 5.

II. STABILITY ANALYSIS OF THE DIFFERENCE EQUATIONS

A. Introduction

In the present state of the art of numerical calculation by finite difference methods, the greatest difficulty is to find accurate and stable approximations. Our numerical results depend crucially on the mathematical tools we use, and our solutions can be meaningful only to the extent that the shortcomings of the approximation are taken into account. It is for these reasons that we will devote a preliminary section to this aspect of computation.

In the case of nonlinear partial differential equations, convergence proofs for numerical approximations are essentially nonexistent. There are no means of making exact error estimates. We must depend upon our numerical experiments to give insight into these matters. Typically, one would assume that a truncation analysis, whereby one examines the order of the approximation in a power series, would lead to some insight. This technique is frequently used but often leads to disappointment because either the expressions become too complicated to analyze or the significance of yet higher-order terms are crucial to full recognition of errors that are present. Here we shall present a different approach by which we can make good use of the computer to aid our understanding. The approach is essentially a wave analysis and is an extension of the simpler notion of obtaining the conditions for stability.

B. Stability Analysis of the Heat Conduction Equation

Consider the heat conduction equation (diffusion of vorticity)

$$\frac{\partial \omega}{\partial \tau} = \nu \, \frac{\partial^2 \omega}{\partial x^2} \tag{2.1}$$

This equation has normalized solutions of the type

$$\omega = re^{ikx} \qquad (2.2)$$

where

$$r = e^{-\nu k^2 t} \qquad (2.3)$$

Here $k = 2\pi/\lambda$ is the wave number, and we note that r is an amplitude factor of the wave. As we expect from our previous discussion of the Laplacian, the wave (2.2) damps and is seen to damp exponentially. The damping depends upon the wave number, and we note that short waves (large wave numbers) damp most rapidly. This follows our intuition since very localized variations in space mean steeper gradients. In general, a distribution of ω is composed of a superposition of waves of the type (2.2).

In the particular case of (2.1), we can get a summed solution (the error integral) that can match a variety of boundary conditions. However, here we are interested only in the abstract notion of the behavior of individual waves. Thus, consider the finite difference approximation

$$\omega_j^{n+1} = \omega_j^{\,n} + \frac{\nu \Delta t}{\Delta x^2} (\omega_{j+1}^n - 2\omega_j^{\,n} + \omega_{j-1}^n) \qquad (2.4)$$

It has the solution

$$\omega_j^{\,n} = r^n e^{ikj\Delta x} \qquad (2.5)$$

where

$$r = 1 + \frac{2\nu \Delta t}{\Delta x^2} (\cos k\Delta x - 1) \qquad (2.6)$$

Note that on the right-hand side of (2.5) j and n should be interpreted algebraically rather than as indices.

There are several things we may do to analyze the behavior of an approximation such as (2.4). First, we should see if there are any restrictions on time step size for a given mesh increment. We cannot permit $|r|$ to exceed 1, because this would give us amplification instead of damping in accord with (2.3). We note from (2.6) that r never exceeds $+1$ and that the most negative value of r occurs when $\cos k\Delta x = -1$. Thus, to prevent amplification from occurring we must have

$$r_{\min} = 1 - \frac{4\nu \Delta t}{\Delta x^2} \geq -1$$

or

$$\frac{\nu \Delta t}{\Delta x^2} \leq \frac{1}{2} \qquad (2.7)$$

We must restrict our calculations to conform to (2.7), otherwise the solution will grow instead of damp. If (2.7) is violated, growth will be most rapid for $k\Delta x = \pi(\lambda = 2\Delta x)$. Now $\lambda = 2\Delta x$ is the shortest representable wave in a finite grid, being specified by only two values per wave length. For $v\Delta t/\Delta x^2 = \frac{1}{2}$, $r = -1$, if $\lambda = 2\Delta x$. Waves of this wavelength will have a 180° phase reversal without damping. This means that even if we are within the stability limit there is certainly something wrong. Incidentally, we can easily see what happens in this case if the stability limit is slightly exceeded. The $\lambda = 2\Delta x$ waves can come into being by roundoff or some other disturbance in our field of numbers. They will then flop 180° at each time step with ever-increasing amplitude. Hence (2.7) should be replaced by

$$\frac{v\Delta t}{\Delta x^2} < \frac{1}{2} \tag{2.8}$$

Further, it would be wise to see how the approximation stands up for all values of $v\Delta t/\Delta x^2 < \frac{1}{2}$ for $\lambda = 2\Delta x$ and also for all longer waves.

We can write (2.3) (replacing t by one time step Δt) as

$$r = \exp\left(-\frac{v\Delta t}{\Delta x^2}(k\Delta x)^2\right) \tag{2.9}$$

Now we can evaluate r in (2.9) and (2.6) as a function of $v\Delta t/\Delta x^2$ and $k\Delta x$ and note the discrepancy of (2.6). The important point is that we can use the computer to evaluate any ideas we may have for a differencing technique. Clearly, we can expect discrepancies, especially for $\lambda = 2\Delta x$, in the method of (2.4). Simply stating the results in this case we find that all waves damp less rapidly than they should for $v\Delta t/\Delta x^2 \leq \frac{1}{8}$ and damp too much for larger values of $v\Delta t/\Delta x^2$. The greatest errors occur at short waves for values of $v\Delta t/\Delta x^2$ approaching the stability limit.

Without the use of a computer to help test our approximation quickly, we would probably have expanded (2.9) in a power series and attempted a comparison of its terms with (2.6). This would be similar to a truncation analysis and would not really be very effective.

C. Stability Analysis of Laplace's Equation

We could now discuss other forms of approximation of (2.1), but instead we shall go on to related considerations. If we now write the Laplacian iteration as

$$\psi_{i,j}^{l+1} = \psi_{i,j}^{l} + \frac{\alpha}{4}(\psi_{i,j-1}^{l} + \psi_{i,j+1}^{l} + \psi_{i-1,j}^{l} + \psi_{i+1,j}^{l} - 4\psi_{i,j}^{l}) \tag{2.10}$$

we may let

$$\alpha = \frac{4\nu \Delta t}{a^2} \qquad (2.11)$$

and note that for a square mesh, where $a = \Delta x = \Delta y$, (2.10) is the two-dimensional counterpart to (2.4). Incidentally, in the two-dimensional case the stability condition (2.8) becomes

$$\frac{\nu \Delta t}{a^2} < \frac{1}{4} \qquad (2.12)$$

Note now that (2.10) with $\alpha = 1$ is our averaging process for obtaining the iterated solution of Laplace's equation. Thus, with a somewhat different interpretation we can consider the meaning of α as an extrapolation parameter.

The iteration process (2.10) is repeated until

$$| \psi^{l+1} - \psi^l | < \varepsilon$$

where ε is an appropriately small number. We are in trouble with (2.10) for $\alpha = 1$ (the stability limit) because there are circumstances (with waves of $\lambda = 2\Delta x$) for which we would never achieve convergence. Therefore, it is apparent that the stability condition is also a condition for convergence. It is not difficult to see that (2.10) does not lend itself to extrapolation (i.e., $\alpha > 1$).

Consider the case $r < 0$. This means a flopping of values in sign with each incrementing of l (if $\lambda = 2\Delta x$). While this is wrong in the heat equation, in the iteration process it may be an advantage. Since local convergence is faster by a process of overshoot, provided there is not excessive overshoot (instability), then this behavior can enhance convergence. This is particularly true since, in the absence of such an overshoot process, the solution is approached asymptotically and the last bit of accuracy will be hard to reach.

What we need for more rapid convergence is a form that will allow us larger α's without causing instability. In other words, we can sacrifice the accuracy sought in the heat equation to produce the overshoot effect that will give more rapid convergence. The final converged solution is our main interest rather than the value of a single iterate or time step as is true in the heat equation.

In the iteration process new values of some of the quantities are already known, so we can revise (2.10) to read

$$\psi_{i,j}^{l+1} = \psi_{i,j}^l + \frac{\alpha}{4} (\psi_{i,j-1}^{l+1} + \psi_{i,j+1}^l + \psi_{i-1,j}^{l+1} + \psi_{i+1,j}^l - 4\psi_{i,j}^l) \qquad (2.13)$$

We now obtain

$$r = \frac{(1 - \alpha) + (\alpha/2) \cos k\Delta x}{1 - (\alpha/2) \cos k\Delta x}$$
(2.14)

Note that for $\alpha = 2$, r is equal to -1 independent of $k\Delta x$ and that all values for $\alpha < 2$ are stable. Overshoot can be regulated so that short waves will overshoot optimally for rapid convergence. A rough guess technique would be to select some mean wavelength expected in solution and require $r \to 0$ for these waves. Under these circumstances all shorter waves will overshoot to enhance the convergence of longer waves. This suggests that larger α's should be used for finer grids. In general a more sophisticated approach must be taken to optimize the convergence rate.

We quickly lose interest in such optimization techniques, however, if they hinge on boundary conditions. The speed gained is not sufficient to justify the efforts required except in special circumstances where many identical types of calculations are to be performed. Further, there may well be better methods than (2.13) that should be investigated both as iterating methods for Laplace's equation and for higher accuracy in the heat equation. In addition, there are Fourier transform techniques that are "one shot" methods. They give accuracy to machine roundoff and are applicable to several varieties of boundary conditions.

D. Stability Analysis of the Nonlinear Convective Equation

Let us now investigate the nonlinear convective equations. In general one would like to find the stability of the complete set of equations, but an overall wave analysis becomes very complicated. As we shall see, most convective approximations contain a diffusive effect that should not be present. This confuses the issue in an overall analysis of the complete equation.

Perhaps the most instructive approach to development of difference approximations for the convective terms is to regard the dependent variable as a physical quantity carried by particles. This is strictly true in the case for pure convection. Further, such an interpretation leads us to conclude that convective difference equations are simply interpolation formulas that enable us to see a whole host of possibilities that may be tried.

Experience has shown that it is almost a general law that we cannot take time steps that would lead to particle motions of more than one mesh distance in one time step. That is, we must have

$$u\Delta t < \Delta x$$
(2.15)

If (2.15) is not satisfied, then we are making use of mesh point information that amounts to extrapolation relative to the particle's nearest neighbor spatial values so that, at the least, accuracy must suffer. Often the stability analysis will show us that under circumstances where (2.15) is not satisfied we will have numerical instability similar to that encountered with the heat equation. In other instances stability may be indicated for all time step sizes but we are faced with a damping of values and with phase errors or changes of shape of a given distribution that are strictly numerical in origin.

Let us consider the simplest case in which we make use of only two neighbor values in developing a difference approximation in one dimension. We may write

$$\frac{\partial \omega}{\partial t} = -u \frac{\partial \omega}{\partial x} \rightarrow \frac{\omega_j^{n+1} - \omega_j^{n}}{\Delta t} \approx u_{j-\frac{1}{2}}^{n} \frac{\omega_{j-1}^{n} - \omega_j^{n}}{\Delta x} \qquad (2.16)$$

where $\omega(x, t) = \omega(j\Delta x, n\Delta t)$. We may interpret this in the $x - t$ plane as an upstream difference and require it to apply only for $u_{j-\frac{1}{2}} > 0$. Consider a particle whose location is between $j - 1$ and j such that $(j - 1)\Delta x \leq x_p \leq j\Delta x$ at time $t = n\Delta t$. The particular particle of interest is that one for which $x_p = j\Delta x$ at time $t = (n + 1)\Delta t$. By distance interpolation we may write for the functional value at the particle

$$\omega_p = \frac{j\Delta x - x_p}{\Delta x} \omega_{j-1}^{n} + \frac{x_p - (j - 1)\Delta x}{\Delta x} \omega_j^{n} \qquad (2.17)$$

so that if $x_p = j\Delta x$, $\omega_p = \omega_j$, and if $x_p = (j - 1)\Delta x$, then $\omega_p = \omega_{j-1}$.

If we now make the estimate $u_{j-\frac{1}{2}}^{n} \Delta t = j\Delta x - x_p$, we can say $\omega_j^{n+1} = \omega_p$ and write

$$\omega_j^{n+1} = \omega_j^{n} + \frac{u_{j-\frac{1}{2}}^{n} \Delta t}{\Delta x} (\omega_{j-1}^{n} - \omega_j^{n}) \qquad (2.18)$$

Thus in (2.17) we see the simple interpolative character of the difference approximation. Clearly we can apply the same approach to any number of space points of our mesh to obtain higher order approximations.

Now, as before, let us consider wave solutions of (2.18) of the form (2.5), assuming u to be constant. Then

$$r = 1 + \frac{u\Delta t}{\Delta x} (e^{-ik\Delta x} - 1)$$

or

$$r = 1 + \frac{u\Delta t}{\Delta x} (\cos k\Delta x - 1) - i \frac{u\Delta t}{\Delta x} \sin k\Delta x \qquad (2.19)$$

Note the similarity of (2.19) to (2.6) with the exception that we now have an imaginary part. The fact that r is complex should have been expected because we know that the spatial waves must shift in phase relative to their original distribution if the convective velocity is nonzero. Of course, if $u = 0$, $r = 1$; our spatial solution (2.5) is unmodified. Similarly, if $u\Delta t/\Delta x = 1$, $r = 1$ and our spatial solution has remained unmodified but is shifted one position Δx downstream for each time step Δt. That is,

$$\omega_j{}^n = e^{-ik\Delta x} e^{ikj\Delta x} = e^{ik(j-1)\Delta x}$$

Clearly, if this were the only concern we could propagate any distribution beautifully from mesh point to mesh point and, as our wave analysis indicates, all waves would move appropriately under (2.18) so that the superposition would give us the propagated distribution without any distortion.

The most obvious drawback to this approach is that our calculations are nonlinear since u is not constant in time or in space. Thus it becomes necessary to interpolate between values. This brings in many deficiencies beyond the assumption that u is a local estimate of the particle velocity of interest. Primarily, we must again require that r be bounded. Because r is complex the modulus of r must be less than 1:

$$r\bar{r} \leq 1 \tag{2.20}$$

That is, r must lie within the unit circle of the complex plane. Let us consider the shortest possible mesh wavelength $k\Delta x = \pi$. Then if $\alpha = u\Delta t/\Delta x$,

$$r = 1 - 2\alpha \tag{2.21}$$

Note that r is real for this wavelength. We already expect difficulty as a consequence. If we require, as before, that $r = 1 - 2\alpha \geq -1$, then $\alpha \leq 1$ and (2.15) applies. For completeness we would have to evaluate (2.20) to ascertain if other waves may be unstable. Simply stating the results we find that the wave $k\Delta x = \pi$ is the one of most concern. If (2.15) is satisfied, then a stable calculation will result.

Note from (2.21) that $r(\alpha)$ does not remain unity. In fact $r = 0$ if $\alpha = \frac{1}{2}$. This means a ripple wave of $k\Delta x = \pi$ will vanish in one time step if $\alpha = \frac{1}{2}$. Our numerical approximation thus has a damping property that causes short waves to become long waves. Here we have noted an extreme special case.

We also find for (2.18) that damping occurs for $0 < \alpha < 1$ with maximum at $\alpha \sim \frac{1}{2}$ for all waves. The damping diminishes rapidly with wavelength so that long waves have essentially no damping. This property becomes an important concern in choosing mesh size. The method given in (2.18) would be rejected on this basis because the grid size required is prohibitively small.

Note also from (2.21) that if $\alpha = 1$, then $r = -1$ and a ripple of $k\Delta x = \pi$ would be maintained correctly. Almost as if by accident, for $\alpha < \frac{1}{2}$ the wave is stationary while for $\alpha > \frac{1}{2}$ the wave jumps to the next grid point.

For other waves the behavior is smoother since they now have an imaginary component. The phase shift should be

$$-\Phi_0 = 2\pi \frac{u\Delta t}{\lambda} = \alpha k \Delta x \tag{2.22}$$

whereas (2.18) gives a phase shift

$$\Phi = \tan^{-1} \frac{I(r)}{R(r)} \tag{2.23}$$

where R and I are, respectively, the real and imaginary parts of r. We may define the error in phase as

$$\Delta\Phi = \Phi_0 - \Phi \tag{2.24}$$

Again stating the results we find that (2.18) has a lagging phase error for $0 < \alpha < \frac{1}{2}$ that results in a distribution not moving as rapidly as it should. If $\frac{1}{2} < \alpha < 1$, we have a leading phase error and the distribution will move more rapidly than it should. The error is largest at $\alpha = \frac{1}{4}$ and $\alpha = \frac{3}{4}$. It also varies with wavelength, being largest for short waves and diminishing rapidly as the wavelength increases so that long waves have essentially no error.

The above covers the basic ideas that must go into the design of a difference method for fluid flow. There are certain severe drawbacks mentioned that still have not been satisfactorily resolved. However, as our understanding grows, we can generate any number of difference methods. With sufficient ingenuity in their selection and use we can look forward with optimism to what will ultimately be solvable.

E. Suggested Reading

1. R. D. Richtmyer, *Difference Methods for Initial Value Problems*, Interscience Publishers, Inc., New York, 1957, Chapters 1 and 6.
2. W. P. Crowley, *J. Comp. Phys.* **1**, 471 (1967).
3. J. E. Fromm, *J. Comp. Phys.* **3**, 176 (1968).

III. APPLICATIONS OF THE NUMERICAL PROGRAM FOR INCOMPRESSIBLE FLOW

A. Introduction

In this chapter a finite difference method for calculating time-dependent incompressible fluid flows is described and results of the technique are illustrated for several applications. The objective is to show the diversity of problems that become tractable through the numerical method rather than to develop a complete treatment in any specific area. Included are solutions of Karman vortex street flows, along with examples which involve the transfer of heat in wakes and results from a study of the Benard problem. The examples given are results that were obtained shortly after the development of the numerical method at the Los Alamos Scientific Laboratory in 1963. There are a number of differences in the computational procedures from those given in Section II. The basic content of this section was first published in the *Proceedings of the Seventh Symposium on Advanced Problems and Methods in Fluid Dynamics, Polish Academy of Sciences.*

One can foresee in the coming years increased use of finite difference methods in the solution of engineering problems and in fundamental studies of the behavior of fluids. Now, however, we are only in the very early stages of this work.

In the differential equations the limiting assumptions of linear behavior, incompressibility, continuum properties, and simplifications of boundary properties are a part of our language. In the case of the difference equations, we also are faced with the limiting properties of stability and errors of truncation. These limitations vary over the range of flow parameters, and clear-cut answers on their restraining effects as yet are unavailable. We must depend upon classical problems that have a broad coverage in the literature to provide us with a basis for comparison with our numerical results.

Ingenuity in the use of finite difference methods can bring us knowledge in many areas and often assist in bridging the gap between theory and experiment.

B. The Differential Equations of Fluid Flow

Consider the basic equations of energy, momentum, and mass conservation in a form in which the fluid properties are constant. These may be written as follows:

Energy:

$$\varrho\left(\frac{\partial}{\partial t} + \mathbf{u} \cdot \nabla\right)I = k\nabla^2 T - P\nabla \cdot \mathbf{u} + \lambda(\nabla \cdot \mathbf{u})^2 + \mu\Phi \qquad (3.1)$$

Momentum:

$$\varrho\left(\frac{\partial}{\partial t} + \mathbf{u} \cdot \nabla\right)\mathbf{u} = \varrho\mathbf{g} - \nabla P + (\lambda + \mu)\nabla(\nabla \cdot \mathbf{u}) + \mu\nabla^2\mathbf{u} \qquad (3.2)$$

Mass:

$$\left(\frac{\partial}{\partial t} + \mathbf{u} \cdot \nabla\right)\varrho = -\varrho(\nabla \cdot \mathbf{u}) \qquad (3.3)$$

In (3.1) I is the internal energy per unit mass, T is the temperature, P is the pressure, $\mathbf{u}$ is the velocity, ϱ the density, and Φ the frictional heating. k, μ, and λ are the thermal conductivity and the coefficients of first and second viscosity. In addition, in (3.2) we have $\mathbf{g}$, a body force.

The examples considered here are of flows in which the density variations are small and can be neglected in the sense of the Boussinesq approximation. That is, the density variation is neglected in all terms except that involving the body force $\mathbf{g}$.[1] In this case we take

$$\varrho = \varrho_0 + \delta\varrho = \varrho_0 - \varrho_0\alpha(T - T_0)$$

where α is the volume expansion coefficient of the fluid.

We shall neglect viscous heating (Φ) in the present studies. Also, we shall take $I = c_v T$, where c_v is the heat capacity at constant volume. Grouping the constant fluid properties in the form

$$\varkappa = \frac{k}{\varrho c_v} \qquad \text{and} \qquad \nu = \frac{\mu}{\varrho}$$

we may rewrite (3.1) through (3.3) as

$$\left(\frac{\partial}{\partial t} + \mathbf{u} \cdot \nabla\right)T = \varkappa\nabla^2 T \qquad (3.4)$$

$$\left(\frac{\partial}{\partial t} + \mathbf{u} \cdot \nabla\right)\mathbf{u} = \mathbf{g} - \alpha(T - T_0)\mathbf{g} - \frac{\nabla P}{\varrho} + \nu\nabla^2\mathbf{u} \qquad (3.5)$$

$$\nabla \cdot \mathbf{u} = 0 \qquad (3.6)$$

The vorticity is defined by

$$\boldsymbol{\omega} = \nabla \times \mathbf{u} \qquad (3.7)$$

(3.5) may be expressed in terms of vorticity in the form

$$\left(\frac{\partial}{\partial t} + \mathbf{u} \cdot \nabla\right)\boldsymbol{\omega} = -\alpha \nabla T x \mathbf{g} + \nu \nabla^2 \omega \qquad (3.8)$$

Equations (3.4), (3.6), (3.7), and (3.8) can be used to establish a set of two-dimensional equations that then can be expressed in finite difference form.

In two dimensions the zero divergence of velocity (3.6) permits us to make use of a stream function ψ defined by the equations

$$u \equiv \frac{\partial \psi}{\partial y} \quad \text{and} \quad v \equiv -\frac{\partial \psi}{\partial x} \qquad (3.9)$$

in a rectangular coordinate system. Through (3.7) the mass conservation equation leads to the relation

$$\frac{\partial^2 \psi}{\partial x^2} + \frac{\partial^2 \psi}{\partial y^2} = -\omega \qquad (3.10)$$

The vorticity has only one component in the two-dimensional case and is hence treated as a scalar.

If $\mathbf{g}$ is taken as a constant body force (gravity) directed in the negative y direction, the momentum equation (3.8) becomes

$$\frac{\partial \omega}{\partial t} + \frac{\partial u\omega}{\partial x} + \frac{\partial v\omega}{\partial y} = \alpha g \frac{\partial T}{\partial x} + \nu\left(\frac{\partial^2 \omega}{\partial x^2} + \frac{\partial^2 \omega}{\partial y^2}\right) \qquad (3.11)$$

Finally, the energy equation (3.4) in two-dimensional form becomes

$$\frac{\partial T}{\partial t} + \frac{\partial uT}{\partial x} + \frac{\partial vT}{\partial y} = \varkappa\left(\frac{\partial^2 T}{\partial x^2} + \frac{\partial^2 T}{\partial y^2}\right) \qquad (3.12)$$

C. The Difference Equations

To derive a set of finite difference equations, corresponding to differential equations (3.9) through (3.12), we prescribe a mesh of cells through which the fluid flows. In Fig. 2 we give one such prescription showing where the variables may be defined for convenient differencing.

A square mesh with side length a is given in Fig. 2, but a rectangular mesh where Δx is not equal to Δy can be used. The designation of velocities at half points along elements of the cells is important for satisfying conservation properties. We consider the cell as a volume element of fluid and

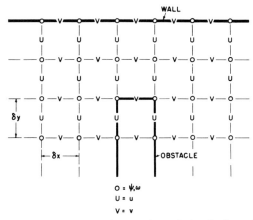

Fig. 2. A representative section of the lattice of points used in the finite difference calculation.

prescribe mass conservation by

$$\frac{u_{i+1,j+\frac{1}{2}} - u_{i,j+\frac{1}{2}}}{\Delta x} + \frac{v_{i+\frac{1}{2},j+1} - v_{i+\frac{1}{2},j}}{\Delta y} = 0$$

When the velocities are given by

$$u_{i,j+\frac{1}{2}} = \frac{\psi_{i,j+1} - \psi_{i,j}}{\Delta y} \quad \text{and} \quad v_{i+\frac{1}{2},j} = \frac{\psi_{i,j} - \psi_{i+1,j}}{\Delta x} \tag{3.13}$$

the continuity equation is automatically satisfied in difference form.

Further, the continuity condition is satisfied for groups of four cells in the form

$$\frac{u_{i+1,j} - u_{i-1,j}}{2\Delta x} + \frac{v_{i,j+1} - v_{i,j-1}}{2\Delta y} = 0$$

if we prescribe, e.g.,

$$u_{i+1,j} = \frac{\psi_{i+1,j+1} - \psi_{i+1,j-1}}{2\Delta y} = \frac{u_{i+1,j+\frac{1}{2}} + u_{i+1,j-\frac{1}{2}}}{2}$$

Also, if the convection of energy is given in the form

$$\frac{(uT)_{i+1,j} - (uT)_{i-1,j}}{2\Delta x} + \frac{(vT)_{i,j+1} - (vT)_{i,j-1}}{2\Delta y} \tag{3.14}$$

no extraneous source of sinks of heat will result. This is true both because of the manner of centering the velocities and because of the use of differences of products of the variables.

An analogous form to (3.14) above also may be used for the transport of vorticity. The vorticity at the point i, j may be defined in terms of nearest neighbor velocities as

$$\omega_{i,j} \equiv \frac{u_{i,j-\frac{1}{2}} - u_{i,j+\frac{1}{2}}}{\Delta y} + \frac{v_{i+\frac{1}{2},j} - v_{i-\frac{1}{2},j}}{\Delta x}$$

Using (3.13) above,

$$\omega_{i,j} = \frac{(\psi_{i,j} - \psi_{i,j-1}) - (\psi_{i,j+1} - \psi_{i,j})}{\Delta y^2} + \frac{(\psi_{i,j} - \psi_{i+1,j}) - (\psi_{i-1,j} - \psi_{i,j})}{\Delta x^2}$$

An explicit differencing scheme with centered time differences is used in this formulation. A variable value at the current time is designated by a superscript n. Backward and forward time values are labeled with superscripts $n - 1$ and $n + 1$, respectively. Thus we would have, e.g.,

$$\frac{T_{i,j}^{n+1} - T_{i,j}^{n-1}}{2\Delta t} = f(u^n, v^n, T^n)$$

However, for numerical stability it is necessary to use a somewhat special form for the conduction terms, also involving backward and forward time values.[2] In these terms we use the Dufort–Frankel[3] method, which makes the approximation

$$\overline{T_{i,j}^n} = \frac{T_{i,j}^{n+1} + T_{i,j}^{n-1}}{2}$$

The two-dimensional conduction terms are

$$\frac{(T_{i,j+1}^n - \overline{T_{i,j}^n}) - (\overline{T_{i,j}^n} - T_{i,j-1}^n)}{\Delta y^2} + \frac{(T_{i+1,j}^n - \overline{T_{i,j}^n}) - (\overline{T_{i,j}^n} - T_{i-1,j}^n)}{\Delta x^2}$$

It is convenient for variable mesh specification to introduce a factor $f = \Delta x/\Delta y$. Then if $\Delta y = a$, $\Delta x = fa$. The complete set of difference equations becomes

(a) Energy conservation

$$T_{i,j}^{n+1} = \left[1 \middle/ \left(1 + \frac{2\varkappa\delta t}{a^2}\frac{f^2 + 1}{f^2}\right)\right]\left\{T_{i,j}^{n-1} - \frac{\delta t}{a}\left[\frac{(uT)_{i+1,j}^n - (uT)_{i-1,j}^n}{f}\right.\right.$$

$$+ (vT)_{i,j+1}^n - (vT)_{i,j-1}^n\right] + \frac{2\varkappa\delta t}{a^2}\left(\frac{T_{i+1,j}^n + T_{i-1,j}^n}{f^2}\right.$$

$$\left.\left. + T_{i,j+1}^n + T_{i,j-1}^n - \frac{f^2 + 1}{f^2}T_{i,j}^{n-1}\right)\right\} \qquad (3.15)$$

(b) Vorticity

$$\omega_{i,j}^{n+1} = \left[1\Big/\left(1 + \frac{2\nu\delta t}{a^2}\,\frac{f^2+1}{f^2}\right)\right]\left\{\omega_{i,j}^{n-1} - \frac{\delta t}{a}\left[\frac{(u\omega)_{i+1,j}^n - (u\omega)_{i-1,j}^n}{f}\right.\right.$$

$$+ (v\omega)_{i,j+1}^n - (v\omega)_{i,j-1}^n\Big] + \frac{\alpha g\delta t}{fa}\,(T_{i+1,j}^{n+1} - T_{i-1,j}^{n+1})$$

$$+ \frac{2\nu\delta t}{a^2}\left(\frac{\omega_{i+1,j}^n + \omega_{i-1,j}^n}{f^2} + \omega_{i,j+1}^n - \omega_{i,j-1}^n - \frac{f^2+1}{f^2}\,\omega_{i,j}^{n-1}\right)\right\}$$

$$(3.16)$$

(c) Stream function

$$\psi_{i,j} = \frac{1}{2(f^2+1)}\,(\psi_{i+1,j} + \psi_{i-1,j}) + \frac{f^2}{2(f^2+1)}\,(\psi_{i,j+1} + \psi_{i,j-1} + a^2\omega_{i,j})$$

$$(3.17)$$

In Eq. (3.16) we arbitrarily chose to write the temperature gradient term using advanced times since these are available if the temperature is computed first.

The order of calculation, assuming known values throughout the mesh at times n and $n-1$, is that of advancing the time to t^{n+1} for the temperature field, then the vorticity field, following this with a simultaneous solution of the stream function field with the new vorticities as source terms in Poisson's equation.

The time step size $\varDelta t$ must in general be restricted in size both for stability and accuracy.[4] The criteria used are such that

$$\frac{u_{max}\delta t}{a} < 1 \quad \text{and} \quad \frac{\nu\delta t}{a^2} < \frac{1}{4}$$

In the first of these a factor less than 1 is chosen to avoid hazards of higher-order instabilities arising from operating near the limits of linear stability.

In (3.17) we do not indicate an iteration procedure but latest values are always used. These latest values of course depend upon the direction in which one sweeps the field of values. A simple criterion to determine adequate convergence of Eq. (3.17) is needed. Experience shows that a satisfactory criterion is that

$$\psi^{h+1} - \psi^h < 0.0002 \qquad (3.18)$$

be satisfied for all points of the mesh, where h and $h+1$ designate successive approximations for ψ having a mean value of unity.

More rapidly convergent forms of successive approximation can be derived where many repetitions of a given type of solution with simple variations of a parameter are made. In such cases the exact method by Hockney[5] is recommended. The literature relating to other methods is cited in his paper.

D. Karman Vortex Street Flows

The first problem studied with our numerical program was that of wakes in a situation where asymmetries in the flow could occur in a natural way as the solution progressed in time. Prior to this time, solutions of this type were restricted to symmetric wakes either because steady solutions were sought[6,7] or, in more recent studies, because of limitations in the capacity of the computer.[8]

Figure 3 shows an early result compared qualitatively with the experimental result obtained by Thom. The numerical solution was a mock-up

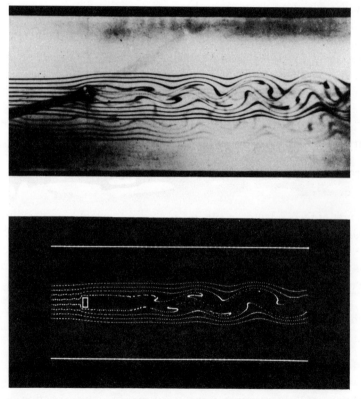

Fig. 3. Experimental and numerical streakline comparison. Experimental result by Thom.[6]

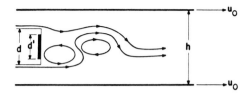

Fig. 4. Geometry of forced heat convection problem.

of a rectangular body being towed through a channel, but the calculation was performed in a reference frame in which the body was stationary. The walls were specified to be moving no-slip walls with the boundary conditions

$$\omega_{i,0} = \frac{u_0 - u_{i+\frac{1}{2}}}{\Delta y} \quad \text{and} \quad \psi_{i,0} = \psi_{i,1} - u_0 \Delta y + (\Delta y)^2 \omega_{i,0} \quad (3.19)$$

specified at the lower wall, where u_0 is the wall velocity.

An analogous condition on the vorticity was applied at the obstacle, but the stream function was maintained at a constant value there. The initial conditions were those of potential flow, and the lateral boundaries were periodic. A variety of results with these boundary conditions are discussed in Reference 4.

The use of spatial periodicity represented a limitation in these studies since the net flow through the channel diminished in time, causing a variation in the effective Reynolds number. Also, the periodic section was limited in size, thus preventing observation of repeating behavior of the wake flows except for a short interval of time. A closer examination of the wake flow was undertaken, and considerations of forced convection of heat were included. This study was suggested by the work of Sogin.[9]

The illustrative sketch of Fig. 4 shows the geometry of the region of the numerical calculation. Again, an impulsive start condition was used: the potential flow solution was one of uniform outflow at the wall velocity u_0 and uniform inflow above and below the after-section of an obstacle at an appropriate velocity to satisfy continuity. Initially only the plate d' was hot with temperature T_1. The fluid and the remaining interfaces were cool at a temperature T_0.

Suitable boundary conditions of no-slip applied at interfaces for ω, while ψ was constant there. Temperatures at surfaces were as prescribed initially, and the inflow was maintained at a temperature T_0 In addition to the Reynolds number $R = u_0 d/\nu$ flows of this type are characterized by a blockage ratio $B = d/h$ and a Prandtl number $P_r = \nu/\varkappa$.

The uniform inflow was maintained, and hence at the entrance the vorticity was zero except at the boundary points. Thus the configuration did not represent a long obstacle body.

Simple linear extrapolation of ψ, ω, and T were prescribed at the outflow. That is, difference approximations for

$$\frac{\partial v}{\partial x} = \frac{\partial}{\partial x}\,(\nabla^2\psi) = \frac{\partial^2 T}{\partial x^2} = 0$$

were used. Such boundary conditions were justified only in the sense that the very near wake was of interest and the reaction of downstream properties could be ignored. The results showed that these outflow conditions were satisfactory for Reynolds numbers less than 400. However, above $R = 100$ the total problem time was limited by numerical instabilities which developed at the outflow.

In Fig. 5 the results of flow for $R = 100$ are illustrated. The flow deviates from a steady behavior with a periodic oscillation of the eddy pair. A slight imbalance in the eddy sizes leads to a growth of the smaller eddy because the transported vorticity producing the latter is momentarily retained in the near region by the circulations of fluid associated with the larger vortex.

R=100

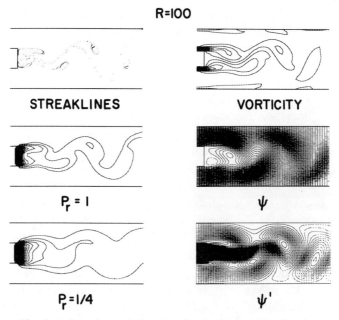

Fig. 5. A late time solution for the forced heat flow study.

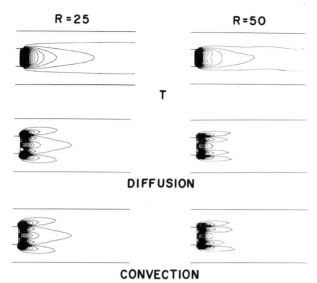

Fig. 6. Contour plots of the instantaneous contributions to the temperature field at the time of the given heat distribution T.

Streaklines, contours of constant vorticity, isotherms, and streamlines are displayed in Fig. 5. The fluid has become heated in the wake, and the effects of reverse flow in producing a high-temperature gradient at the plate d' are evident. In an analogous way to the conduction and convection of heat from the rear plate, the initial distribution of vorticity at the boundaries, corresponding to the potential flow conditions, has here become transported and has diffused to produce the given distribution. A boundary layer at the moving walls develops because the constricted flow exceeds the speed of the moving walls. Further details of variation of R, B, and P_r are given by Harlow and Fromm.[10]

Figure 6 is included to help understand the nonsteady behavior resulting from the imbalance of the instantaneous contributions by the nonlinear convection terms and the conduction terms in their influence on the temperature distribution. In the very near wake a near balance exists. One may identify the sign of the contribution in this near region and then throughout by noting an increase due to conduction at the plate d' with an almost identical region of decrease by convection in the same area.

E. The Benard Problem

As a final example of application of finite difference techniques to nonlinear flows we consider a case where the temperature field reacts back on the flow field. The problem is that of thermal convection in a horizontal fluid layer. The linear instability problem has been solved and is discussed in considerable detail in the treatise by Chandrasekhar.[1] Nonlinear numerical solutions have been obtained by Deardorff[11] and by Fromm.[12]

The flows to be discussed were first observed by Benard and were initially studied analytically by Rayleigh. They are characterized by the parameter

$$R = \frac{\alpha g \, \Delta T h^3}{\varkappa \nu}$$

the Rayleigh number. Here h is the depth of the layer and ΔT is the temperature between the upper and lower bounding surfaces of the layer. Buoyancy is present as a consequence of heating the plate to a temperature ΔT above the upper plate. At $R = R_c$, the critical Rayleigh number, fluid motions first commence at a wave number $k = k_c$. This has been predicted by linear theory and has been observed experimentally. Analytic solutions of the nonlinear problem have been sought by many authors, but of these only the solutions obtained by Kuo[13] are quantitatively correct.

Numerically and analytically the fluid layer of infinite extent is treated by considering a periodic section chosen to have a length corresponding to the wave number k_c. The free boundary problem, i.e., one with free slip horizontal surfaces, is studied. This is a tractable analytic problem and contains a number of the important flow features of real flows.

The initial conditions were those of no flow, but a linear temperature variation between the horizontal surfaces was preestablished. This is a realistic initial condition at low Rayleigh numbers since indeed the instability will not arise until conduction is taking place through the layer.

The initial stratified temperature distribution becomes modified in time if a perturbation is applied. In the numerical calculation the horizontal center line of temperature points is given a very small spatially sinusoidal perturbation. An exponential growth of fluid circulation starts but eventually stops because the redistributed temperature field will not further enhance the circulation. In other words, the nonlinear effects terminate the growth of the linear instability.

The results of Fig. 7 clearly show the validity of Kuo's analysis and also the value of finite difference computation. Slight differences suggest that perhaps some higher order terms are necessary in the analysis of the flows illustrated in Fig. 7 to obtain better agreement at rising columns of heat.

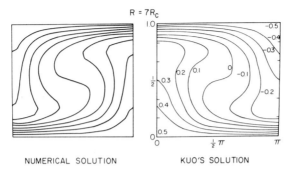

NUMERICAL SOLUTION   KUO'S SOLUTION

Fig. 7. Steady-state solution of the thermal convection problem showing agreement of numerical and analytical results.

We have shown how the finite difference method has been used to
obtain solutions to several problems. It is clear that in all cases the linear
stability problem can readily be extended to an associated nonlinear one
through numerical computation. In addition to such fundamental studies,
much can be done in solving engineering problems of basic interest.

F. References

1. S. Chandrasekhar, *Hydrodynamic and Hydromagnetic Stability*, Oxford University Press, 1961.
2. R. Richtmyer, *Difference Methods for Initial Value Problems*, Interscience Publishers, Inc., New York, 1957.
3. E. Dufort and S. Frankel, Stability conditions in the numerical treatment of parabolic differential equations, *Math. Tables and Other Aids to Computation* **7**, 135 (1953).
4. J. Fromm, A method for computing nonsteady, incompressible, viscous fluid flows, *Los Alamos Scientific Laboratory Report LA 2910*, Los Alamos (1963).
5. R. Hockney, A fast, direct solution of Poisson's equation using Fourier analysis, *Stanford Electronics Laboratory Technical Report No. 0255-1*, Stanford (1964).
6. A. Thom, The flow past circular cylinders at low speeds, *Proc. Roy. Soc.* **A141**, 651 (1933).
7. M. Kawaguti, Numerical solution of the Navier–Stokes equations for the flow around a circular cylinder at Reynolds Number 40, *J. Phys. Soc. Japan* **6**(8), 747–757 (1953).
8. R. Payne, Calculations of unsteady viscous flow past a circular cylinder, *J. Fluid Mech.* **1**(4), 81 (1958).
9. H. Sogin, Heat transfer from the rear of bluff objects to a low speed air stream, *Tulane University Aeronautical Research Laboratories Report No. ARL-62-361* (1962).
10. F. Harlow and J. Fromm, Dynamics and heat transfer in the von Karman wake of a rectangular cylinder, *Phys. Fluids* **7**, 1147 (1964).
11. J. Deardorff, A numerical study of two-dimensional parallel-plate convection, *J. Atmos. Sci.* **21**, 414–438 (1964).

12. J. Fromm, Numerical solutions of the nonlinear equations for a heated fluid layer, *Phys. Fluids* **8**, 1757 (1965).
13. H. Kuo, Solutions of the nonlinear equations of cellular convection and heat transport, *J. Fluid Mech.* **4**(10), 611–634 (1961).

IV. DESCRIPTION OF THE NUMERICAL PROGRAM FOR INCOMPRESSIBLE FLOW

A. Discussion of the Block Diagram

Since our main purpose is to transmit understanding of the numerical methods, we shall include in this section only a skeleton program for computing incompressible flows. We will not discuss the graphics subroutines or the routines to preserve and restart a given problem. Also, only the equations of Section I are used, giving us just the basic flow dynamics in the absence of all but the convective (inertial) and viscous forces. Nevertheless, there is a certain completeness in the given program in that additions may be made as insertions without involving major reprogramming. Figure 8 gives a block diagram of the subroutines called by the mainline program.

We begin the block program by entering DATAIN which, as the name implies, reads input data. The input data include a minimum of input parameters such as grid size, grid aspect ratio, print interval, and Reynolds number. The DATAIN program also is used to consolidate parameters to provide for simplified constants that will later speed the computation. It also determines the size of the initial time step.

The program GEOIN, like DATAIN, reads input data but is specialized to give a flexibility of geometry with a minimum number of data words. The boundary must consist of straight line segments passing through mesh points, but this is the only restriction on the geometry. GEOIN also establishes the initial values for boundary points.

The program GEOPRN gives a symbolic printout of input geometry for checkout purposes. The symbols used characterize the type of boundary condition that is to be imposed (rigid no-slip, rigid free-slip, periodic, etc.).

Before starting the computation it is convenient to introduce an additional program (IGUESS) to establish initial guess values based on already prescribed boundary values. An intelligent guess enables us to speed up the convergence of our initial ψ solution. This initial guess varies from a crude layout of values given by a linear interpolation between boundary values

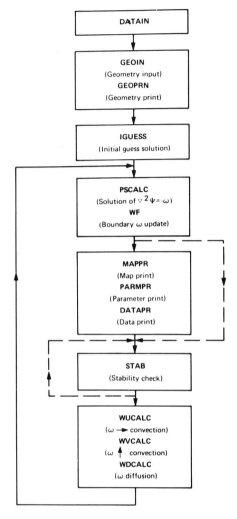

Fig. 8. Block diagram of the numerical program.

all the way to known exact solutions. If some initially specified distribution is to be perturbed, this is also programmed in the subroutine (IGUESS).

The program enters the subroutine PSCALC for actual solution of Poisson's equation. Here, through an iteration (averaging) process, we obtain the stream function distribution. Most often the stream function is the potential flow solution for the given geometry. After obtaining convergence to some specified accuracy we establish boundary values of vorticity at no-slip surfaces. This is done in the subroutine WF.

At this point in the calculation we have a complete solution with a consistent set of numbers for ψ and ω. We may wish to map these numbers or print them. It also may be desirable to include input parameters or functions evaluated from the data for reference or identification relative to the maps or printouts. The program MAPPR is used to print a map of normalized and highly rounded values to give a picture that bears some resemblance to a contour map. A good printer map in some cases may provide a substitute for a contour map. The DATAPR program simply prints out the data values at all grid points in a layout that is geometrically similar to the given maps but usually requires several pages to contain the information. Finally PARMPR serves as a labeling program, providing orientation information for the given maps or printouts.

Basically this is a logical point to terminate the program. However, we have not yet advanced our solutions in time; we have only obtained a consistent initial solution. From this initial solution we now may reevaluate the size of time step required to maintain stability. This is done in the program STAB. If it tells us to reduce the time step that was initially furnished, we can use STAB to reevaluate any constants that are affected by this change. The dotted arrow in Fig. 8 symbolizes a repeated reentry into the program until stability conditions are appropriate for proceeding with the computation.

The next set of subroutines allows for the advancement of the vorticity distribution to correspond to a new time, one time increment (Δt as established by STAB) beyond our previous solution. WUCALC computes the convection of the vorticity distribution in the coordinate direction of the u velocity (x coordinate). Similarly, WVCALC gives the convection associated with the v velocity (y direction). Finally, WDCALC incorporates the diffusion of vorticity that will occur over the time interval. The time-advanced distribution of vorticity will induce a change in the flow so we now must evaluate a new stream function distribution by reentering PSCALC. Each passage through this loop of subroutines amounts to the advancement in time of the solution by a small discrete time interval. We may or may not record data at each time step, but the stability is tested every time step.

B. Discussion of the Program Listings

Now proceed to the actual listings of the programs. We cannot cover them in every detail, but some comments will be useful to those who plan to run the program later. One important point about these listings is that there are no details missing so far as the physics of the calculation is concerned. (If there are any facets of the work that are not clear from the preceding description, they can be studied in their entirety in the programs.)

Because the program has been extracted from a more extensive set of programs, there are more parameters defined in the data lists than are here applicable. The subroutine calls sometimes contain redundant arguments. Generally, the best procedure for learning a program from listings is to draw flow charts, including all detail, for each subroutine. Since such thorough study is best left to the individual, we shall only scan the programs for points that may assist individual study.

Input data involves array size information and data as read by DATAIN and GEOIN. The array size data is given in the first six cards of the mainline program, while the remaining data is included at the end of the program deck. Logical arrays are used to establish boundary condition flags for both ψ and ω fields. Both WPOINT(ω) and PPOINT(ψ) have the following format:

	I	II	III	IV
TRUE	FLUID POINT	SLIP	MOVING	CONTINUATIVE
FALSE	BOUNDARY POINT	NO–SLIP	NOT MOVING	PERIODIC

For a fluid point, IV is applicable if it is at the edge of the array. For boundary points, II and/or III apply with some redundancy (slip and moving have no meaning).

Edge points, EPOINT, make rather wasteful use of space by a similar format:

	I	II	III	IV
TRUE	UPPER	LOWER	RIGHT	LEFT
FALSE				

These labels are gradually being replaced by tests on indices, and some programs will reflect this.

Common storage parameters are identified in the program DATAIN. The call to DATAIN is preceded only by a call to UTSX, a routine for suppressing the print of underflow messages.

A program BLOCK DATA is provided but is not called explicitly. Through FORTRAN it will initialize common storage parameters. DATAIN reads in replacements of these parameters, as necessary, and establishes new constants. The initial time step established is an upper bound for diffusional stability. The diffusional stability limit does not require modification during computation if the kinematic viscosity v is a constant.

The namelist INPUT is read in DATAIN, and the namelist SPECS is read in GEOIN. The latter needs some explanation. The first card gives the variable to which a given subset of cards applies and the number of such cards. The no-slip condition is implied so that (slip) must be specified on this first card if it is to apply. Inflow and outflow boundaries are pre-specified as periodic and must be changed to continuative if this is desired. The inflow and outflow boundaries are the left and right edges of the array but do not include the corner points of the array.

The cards following the SPECS card have the following format:

Columns 1 and 2	00 Horizontal line
	10 Vertical line
	$+1$ Positive slope line
	-1 Negative slope line
Columns 4–9	Initial values to be given ω for boundaries prescribed on this same card (floating point number)
Columns 11–16	Initial values to be given ψ for boundaries prescribed on this same card (floating point number)
Columns 18–23	For additional variable not here discussed
Columns 25–33 Locus # 1	Columns 25, 26, 27—initial x coordinate (integer index value) Columns 28, 29, 30—initial y coordinate (integer index value) Columns 31, 32, 33—terminal x coordinate if horizontal or sloped line Terminal y coordinate if vertical line (integer index value)
Columns 35–43 Locus # 2	Same as Locus # 1 for second line segment
Columns 45–53 Locus # 3	Same as Locus # 1 for third line segment
Columns 55–63 Locus # 4	Same as Locus # 1 for fourth line segment
Columns 65–73 Locus # 5	Same as Locus # 1 for fifth line segment
Columns 75–80	For additional variable not here discussed

Any number of cards may be included, but if specifications overlap, those on the last specification card will be used. These input cards are interpreted by GEOIN to give each mesh point a label that determines its treatment in later subroutines.

PSCALC is the routine that solves Poisson's equation. The arguments allow for an extrapolation parameter value, a number governing the degree of convergence, and an integer specifying a minimum number of iterations to guard against local convergence in the absence of overall convergence. The subroutine PSCALC iterates symmetrically about a central horizontal line and has provisions for extrapolation in time rather than simply using the previous time step solution for a first guess. This is subsidiary to the extrapolation mentioned above and serves to provide an improved guess over the prior solution if rapid time variation is taking place. The last two arguments in the call to PSCALC are provided as addresses for special fields used in connection with reducing indexing computations. This necessitates repetition of the arguments in the PSCALC subroutine call.

The subroutines WUCALC and WVCALC are written in such a form as to force vorticity conservation. A given flux of vorticity is added to a particular cell value, and at the same time this value is subtracted from the adjacent cell. This process guarantees that the difference approximation itself will not be the source of extraneous, numerical vorticity. This mode of programming does not affect the linear stability condition as evaluated in the routine STAB.

The program WUCALC included in the listings constitutes a fourth-order scheme that will take some individual study to understand fully. Since the fourth-order routine uses more neighboring mesh points there is the problem of not having sufficient information next to a boundary when the flow is away from the boundary. We have taken the easiest course, which is to substitute a second-order form (or even a first-order form) as is necessary to permit evaluation. It may be possible to include higher order boundary conditions in some cases, but this has not been explored.

Finally, the routine WDCALC performs the diffusional part of the vorticity transport. The reset of boundary values in the initial part of this program is a matter of convenience because flux transfers from and to boundaries in the previous programs was allowed. These changed values at the boundaries are superfluous and relate to the programming technique. For this reason they must be reset to their original values so as not to affect the diffusional transfer.

In WDCALC the Laplacian is evaluated, in accordance with the given boundary conditions, and the contribution is added to give the final new value to the vorticity distribution. At this stage we reenter PSCALC to get the new flow field corresponding to the new vorticity distribution.

Please note that the included programs are not complete because of space limitations. The excluded programs can either be readily generated by a potential user or, as with WVCALC, an understanding of other parts makes it possible to rebuild the routine through correspondence.

C. Suggested Reading

1. E. I. Organick, *A FORTRAN Primer*, Addison-Wesley Publishing Co., Inc., Palo Alto, 1963.

APPENDIX:

Computer Listings
of the Hydrodynamic Programs

```
C     MAIN PROGRAM
      LOGICAL*1 WPOINT(57, 25,4),                    PPOINT(57, 25,4),
     X          EPOINT(57, 25,4)
      DIMENSION PSP1(57, 25),               WP1(57, 25),
     X          PS  (57, 25),               S  (57, 25)
      IDIMX = 57
      IDIMY=25
      COMMON/INDATA/ A    , F    ,DT,DTP, DTPR, UL, UU, D, VIS, CON,  M   126
     X          HALG, AM, WN, TEST, DTTS, XLL, YLL, XUR, YUR, XGP,X   M   127
     X          YGP, IDIMX, IDIMY, TDIF, APG, TEMP, PRES             C3
     X        /USDATA/ AS, FS, CF, AF, ASFS, TM, ITT, R, AL, BT, C, H,  M   129
     X          TP, TPR  ,CFI
      INTEGER DTP, DTPR, DTTS                                        M   105
      DATA IPRCNT/0/
      INTEGER*2 LITOW/' W'/, LITOP/' P'/
      DATA LITWP1/'WP1 '/, LITTP1/'VR3 '/, LIPSP1/'PSP1'/
      CALL UTSX
      DO 15 J = 1,IDIMY,1
      DO 15 I = 1,IDIMX,1
      DO 10 K = 1,4,1                                                M   202
      WPOINT (I,J,K) = .FALSE.                                       M   205
      PPOINT (I,J,K) = .FALSE.                                       M   207
   10 EPOINT (I,J,K) = .FALSE.                                       M   208
      WP1  (I,J) = 0.0                                               M   209
      PSP1 (I,J) = 0.0
      S    (I,J) = 0.0
   15 PS   (I,J) = 0.0
C     INDICATOR ARRAYS INITIALIZED AND DATA ARRAYS ZEROED.          M   217
      CALL  DATAIN                                                   M   220
      CALL       GEOIN(WP1,S    , PSP1, PSP1, WPOINT, WPOINT, PPOINT,
     X        PPOINT,      EPOINT, IDIMX, IDIMY, &100,&100)
  100 CALL       GEOPRN(WPOINT,EPOINT, LITOW,IDIMX,IDIMY)
      CALL       GEOPRN(PPOINT,EPOINT, LITOP,IDIMX,IDIMY)
      CALL       IGUESS(WP1,PSP1,S   ,WPOINT,PPOINT,WPOINT,EPOINT,
     X     IDIMX,IDIMY)
      IPRCNT = DTPR
  109 CALL       PSCALC(PSP1,PS,S   ,WP1,S  ,PPOINT,PPOINT,
     X     IDIMX,IDIMY,1.5,0.00005,10,PSP1,WP1)
      CALL       WF(WP1,PSP1,WPOINT,PPOINT,IDIMX,IDIMY)
      IPRCNT = IPRCNT + 1
      IF (IPRCNT .LT. DTPR) GO TO 33
      IPRCNT = 0                                                     M   271
      CALL       MAPPR (WP1 ,LITWP1,IDIMX,IDIMY)
      CALL PARMPR                                                    M   274
      CALL       MAPPR (PSP1,LIPSP1,IDIMX,IDIMY)
      CALL DATAPR (WP1 ,LITWP1,IDIMX,IDIMY)
      CALL DATAPR (PSP1,LIPSP1,IDIMX,IDIMY)
   33 CALL       STAB(PSP1, EPOINT, IDIMX, IDIMY,PS ,PS  ,&66,&33)
   66 DO 121 J = 1, IDIMY,1
      DO 121 I = 1, IDIMX,1
  121 S   (I, J) = WP1 (I, J)
      CALL       WUCALC(WP1,S,S   ,S   ,PSP1,PS   ,S ,S ,WPOINT,EPOINT,
     X     IDIMX,IDIMY)
      DO 222 J = 1, IDIMY,1
      DO 222 I = 1, IDIMX,1
      IF (WPOINT(I,J,1)) S    (I,J) = WP1(I,J)
      IF (.NOT.WPOINT(I,J,1)) WP1 (I,J) = S    (I,J)
  222 CONTINUE
      CALL       WVCALC(WP1,S,S   ,S   ,PSP1,PS   ,S ,S ,WPOINT,EPOINT,
     X     IDIMX,IDIMY)
      DO 122 J = 1, IDIMY,1
      DO 122 I = 1, IDIMX,1
      IF (WPOINT(I,J,1)) S    (I,J) = WP1(I,J)
      IF (.NOT.WPOINT(I,J,1)) WP1 (I,J) = S    (I,J)
  122 CONTINUE
      CALL       WDCALC(WP1,S,S   ,S   ,PSP1,PS   ,S ,S ,WPOINT,EPOINT,
     X            IDIMX,IDIMY)
      DO 325 J = 1, IDIMY,1
      DO 325 I = 1, IDIMX,1
      S   (I, J) = PS   (I, J)
  325 PS  (I, J) = PSP1(I, J)
      GO TO 109
      END



      BLOCK DATA                                                    BD   100
      INTEGER DTP, DTPR, DTTS
      LOGICAL TEMP,PRES
      COMMON /INDATA/ A, F, DT, DTP, DTPR, UL, UU, D, VIS, CON, HALG,  BD   120
     X          AM, WN, TEST, DTTS, XLL, YLL, XUR, YUR, XGP, YGP,  BD   121
     X          IXDIM, IYDIM, TDIF,   APG,     TEMP,  PRES,        BD   122
     X          BLANKE, BLANKF, BLANKG, BLANKH, BLANKI, BLANKJ,    BD   123
     X          BLANKK, BLANKL, BLANKM, BLANKN, BLANKO, BLANKP,    BD   124
```

```
      X                    BLANKQ, BLANKR, BLANKS                        BD   125
      X       /USDATA/ AS, FS, CF, AF, ASFS, TM, ITT, R, AL, BT, C, H,   BD   130
      X              TP, TPR, CFI, BLANKT, BLANKU, BLANKV, BLANKW,       BD   131
      X              BLANKX, BLANKY, BLANKZ, BLANKO, BLANK1, BLANK2,     BD   132
      X              BLANK3, BLANK4, BLANK5, BLANK6, BLANK7, BLANK8,     BD   133
      X              BLANK9                                              BD   134
       DATA A/0.0/,F/0.0/,DT/0.0/,DTP/ 0 /,DTPR/ 0 /,UL/0.0/,UU/0.0/,
      X      D/0.0/,VIS/0.0/,CON/0.0/,HALG/0.0/,AM/0.0/,WN/0.0/,TEST/0.0/,
      X      DTTS/ 0 /,XLL/0.0/,YLL/0.0/,XUR/0.0/,YUR/0.0/,XGP/0.0/,
      X      YGP/0.0/,IXDIM/0/,IYDIM/0/,TDIF/0.0/,APG/0.0/,
      X TEMP/.FALSE./,PRES/.FALSE./, BLANKE/0.0/, BLANKF/0.0/,            BD   152
      X      BLANKG/0.0/, BLANKH/0.0/, BLANKI/0.0/, BLANKJ/0.0/,          BD   153
      X      BLANKK/0.0/, BLANKL/0.0/, BLANKM/0.0/, BLANKN/0.0/,          BD   154
      X      BLANKO/0.0/, BLANKP/0.0/, BLANKQ/0.0/, BLANKR/0.0/,          BD   155
      X      BLANKS/0.0/,                                                 BD   156
      X      AS/0.0/,FS/0.0/,CF/0.0/,AF/0.0/,ASFS/0.0/,TM/0.0/,ITT/0/,
      X      R/0.0/,AL/0.0/,BT/0.0/,C/0.0/,H/0.0/,
      X      TP/0.0/, TPR/0.0/, CFI/0.0/, BLANKT/0.0/, BLANKU/0.0/,       BD   161
      X      BLANKV/0.0/, BLANKW/0.0/, BLANKX/0.0/, BLANKY/0.0/,          BD   162
      X      BLANKZ/0.0/, BLANKO/0.0/, BLANK1/0.0/, BLANK2/0.0/,          BD   163
      X      BLANK3/0.0/, BLANK4/0.0/, BLANK5/0.0/, BLANK6/0.0/,          BD   164
      X      BLANK7/0.0/, BLANK8/0.0/, BLANK9/0.0/                        BD   165
       END                                                               BD


      SUBROUTINE  DATAIN
      INTEGER DTP, DTPR, DTTS
      COMMON /INDATA/ A     , F     , DT, DTP, DTPR, UL, UU, D, VIS, CON,
     X                HALG, AM, WN, TEST, DTTS, XLL, YLL, XUR, YUR, XGP,
     X                YGP, IXDIM, IYDIM,TDIF, APG
C          INDATA CONTAINS A BLOCK OF VARIABLES WHICH SPECIFY PROBLEM
C          PARAMETERS.
C               A     IS THE DISTANCE BETWEEN Y COORDINATE MESH POINTS.
C               F     IS THE RATIO OF THE DISTANCE BETWEEN X COORDIN-
C                     ATE MESH POINTS & Y COORDINATE MESH POINTS.
C               DT    IS THE TIME STEP INTERVAL.
C               DTP   IS THE NUMBER OF TIME STEPS BETWEEN PLOTS.
C               DTPR  IS THE NUMBER OF TIME STEPS BETWEEN PRINTS.
C               UL    IS THE X VELOCITY OF THE LOWER BOUNDARY.
C               UU    IS THE X VELOCITY OF THE UPPER BOUNDARY.
C               D     IS THE DIAMETER OF THE OBJECT IN THE FLOW CHAN-
C                     NEL.
C               VIS   IS THE VISCOSITY OF THE FLUID.
C               CON   IS THE THERMAL DIFFUSIVITY.
C               AM    IS THE PETERBATION AMPLITUDE.
C               WN    IS THE NUMBER OF CYCLES IN THE PETERBATION WAVE.
C               DTTS  IS THE NUMBER OF TIME STEPS BETWEEN TAPE DUMPS.
C               XLL   IS THE COORDINATE OF THE LOWER BOUNDARY.
C               YLL   IS THE COORDINATE OF THE RIGHT BOUNDARY.
C               XUR   IS THE COORDINATE OF THE UPPER BOUNDARY.
C               YUR   IS THE COORDINATE OF THE RIGHT BOUNDARY.
C               XGP   IS THE X COORDINATE OF A MESH POINT.
C               YGP   IS THE Y COORDINATE OF A MESH POINT.
C               IXDIM IS THE NUMBER OF MESH POINTS IN THE X DIRECTION.
C               IYDIM IS THE NUMBER OF MESH POINTS IN THE Y DIRECTION
      COMMON /USDATA/ AS, FS, CF, AF, ASFS, TM, ITT,R,AL,BT,C,H,TP ,TPR,
     X  CFI
C          USDATA
C               AS    IS DELTA Y SQUARED.
C               FS    IS RATIO OF (DELTA X TO DELTA Y) SQUARED.
C               CF    IS A MATHEMATICAL CONSTANT USED THROUGHOUT THE
C                     COMPUTATIONS.
C               AF    IS DELTA X.
C               ASFS  IS THE PRODUCT OF AS TIMES FS.
C               TM    IS THE CURRENT PROBLEM TIME.
C               ITT   IS THE NUMBER OF ITERATIONS REQUIRED BEFORE THE
C                     STREAM FUNCTION CONVERGED TO A SOLUTION.
C               R     IS THE REYNOLDS OR RAYLEIGH NUMBER.
      NAMELIST /INPUT/ A,F,DT,DTP,DTPR,UL,UU,D,VIS,CON,HALG,AM,WN,TEST,
     XDTTS, XLL, YLL, XUR, YUR, XGP, YGP, IXDIM, IYDIM,TDIF, APG
      NAMELIST/DATUSE/ R, AL, BT, C, H, ITT, TM, TP, TPR
      READ( 5,INPUT,END=1)
    1 AS = A*A
      FS = F*F
      CF = 2.0*(FS+1.0)
      CFI = 1.0/CF
      AF = A*F
      ASFS = AS*FS
      IF (TEST)13,3,2
   13 DIFC = VIS*DT/AS
      TRAC = UL*DT/A
      GO TO 9
    2 IF  (UL) 21,3,21
```

```
   21 DTU = AF/ABS(UL)
      GO TO 22
    3 DTU = 1000.0
   22 IF (VIS) 5,5,4
    4 DTN = AS/(4.0*VIS)
      GO TO 6
    5 DTN = 1000.0
    6 IF (DTN-DTU) 7,8,8
    7 DT = DT*DTN
      GO TO 9
    8 DT = DT*DTU
    9 DTT = 1.0
      IF (DT-DTT) 20,40,30
   20 DTT = DTT/2.0
      IF (DT-DTT) 20,40,35
   35 DT = DTT
      GO TO 40
   30 DTT = DTT*2.0
      IF (DT-DTT) 39,40,30
   39 DT = DTT/2.0
   40 CONTINUE
      TM = 0.0
      AL = DT/AF
      BT = DT/A
      C = VIS*DT/AS
      H = CON*DT/AS
      IF (TEST) 10,11,12
C        NUMERICAL INSTABILITY TEST
   10 R = 0.0
      GO TO 25
   11 IF (CON) 24,24,41
   41 IF (VIS) 24,24,31
   31 R = 2.0*HALG*TDIF              *D*D*D/(VIS*CON)
      GO TO 25
   12 IF (VIS) 24,24,32
   32 R = UL*D/VIS
   25 WRITE (6,INPUT)
      WRITE (6,DATUSE)
   19 RETURN
   24 WRITE (6,240)
  240 FORMAT(' ','YOU HAVE ASKED FOR AN R WITH VIS OR CON = 0')
      GO TO 19
      END


      SUBROUTINE GEOIN(WP1, TP1, PSP1, PRP1, WPOINT, TPOINT, PPOINT,   GI   100
     X                 PRPONT, EPOINT, IXDIM, IYDIM, *, *)            GI   100
C     SCALAR VARIABLES                                                GI   100
C                                                                     GI   100
C        PS1, OMEGA, TEMP, CONT, SLIP & MOVE ARE LOGICAL VARIABLES OF GI   100
C              LENGTH 1. THEY ARE SET FALSE BY THE PROGRAM. AN ATTEMPTGI   100
C              IS THEN MADE TO READ AT LEAST ONE OF THE FIRST FOUR AS GI   100
C              TRUE BY A NAMELIST READ. THEY ARE USED TO INDICATE THATGI   100
C              A PARTICULAR VALUE IS TO BE READ FROM FOLLOWING INPUT   GI   100
C              DATA.                                                  GI   100
C        ICARD IS A STANDARD LENGTH INTEGER VARIABLE USED IN THE NAME-GI   100
C              LIST READ TO INDICATE HOW MANY FORMATTED DATA CARDS     GI   100
C              WILL FOLLOW THE NAMELIST DATA CARDS.                   GI   100
C        INDERR IS AN INTEGER VARIABLE USED TO INDICATE THAT ERRORS   GI   100
C              HAVE OCCURED ON DATA INPUT.                            GI   100
C        I &J  ARE INTEGER VARIABLES USED AS POINTERS WHEN OPERATING  GI   100
C              ON THE ARRAYS IN LABELED COMMON, TIMEP1 & INDIC         GI   100
C        L     IS AN INTEGER VARIABLE USED AS A POINTER WHEN OPERATING GI   100
C              ON THE ARRAYS INDX, INDY, INDE                         GI   100
C        ICOUNT IS AN INTEGER COUNTER USED BY A DO LOOP TO COUNT      GI   100
C              THROUGH THE ICARD CARDS FOLLOWING A NAMELIST READ.     GI   100
C        ISLOPE IS AN INTEGER VARIABLE WHICH INDICATES THE SLOPE OF THEGI   100
C              BOUNDARIES DEFINED BY THE CARD.                        GI   100
C              ISLOPE = -1 THE BOUNDARY IS DOWNWARD DIAGONALLY.       GI   100
C                         THROUGH VERTICIES OF THE ARRAY.             GI   100
C                     =  0 THE BOUNDARY IS HORIZONTAL.                GI   100
C                     =  1 THE BOUNDARY IS UPWARD DIAGONALLY.         GI   100
C                     = 10 THE BOUNDARY IS VERTICAL.                  GI   100
C        PVAL, WVAL & TVAL ARE FLOATING POINT VARIABLES READ IN AND   GI   100
C              ASSIGNED TO THOSE ARRAY LOCATIONS ON THE BOUNDARY BEINGGI   100
C              DEFINED. THE CORRESPONDENCE IS PVAL WILL BE ASSIGNED    GI   100
C              IN PSP1, WVAL IN WP1 & TVAL IN TP1.                    GI   100
C        IPOINT IS AN INTEGER POINTER USED BY COMPUTED GO TO'S. ITS   GI   100
C              VALUE IS DETERMINED BY THE SLOPE OF THE BOUNDARY.      GI   100
C        ILOW & IUPPER ARE INTEGER LIMITS ON A DO LOOP. THEIR VALUE IS GI   100
C              DETERMINED BY THE BOUNDARY COORDINATES & SLOPE.        GI   100
C        INDEX IS AN INTEGER COUNTER USED BY THE DO LOOP ASSOCIATED   GI   100
C              WITH ILOW & IUPPER.                                    GI   100
C        M     IS AN COUNTER USED TO OPERATE ON ARRAYS INDX, INDY ANDGI   100
```

```
C                        INDE IN THE WRITE STATEMENT ASSOCIATED WITH ERRORS IN   GI  100
C                        THE VALUES READ INTO THESE ARRAYS.                       GI  100
C              LITOP, LITOW, LITOT ARE INTEGER VARIABLES OF LENGTH 2 WHICH        GI  100
C                        ARE USED AS ALPHABETIC CHARACTERS & PASSED TO GEPORN     GI  100
C                        FOR TITLING PURPOSES                                     GI  100
        DIMENSION INDX(5),INDY(5), INDE(5)                                        GI  101
        LOGICAL*1 OMEGA, PSI, TEMP, CONT, SLIP, MOVE, PRES                        GI  102
C        ARRAYS                                                                   GI  100
C              INIX, INDY & INDE ARE ARRAYS WITH DIMENSION 5 INTO WHICH ARE       GI  100
C                        READ COORDINATE VALUES OF BOUNDARIES.                    GI  100
C        NAMELISTS                                                                GI  100
C              SPECS   IS THE NAME OF A NAMELIST USED TO READ DATA.               GI  100
        LOGICAL*1 WPOINT, TPOINT, PPOINT, PRPONT, EPOINT                          GI  105
        DIMENSION WP1 (IXDIM, IYDIM),       TP1 (IXDIM, IYDIM),                   GI  106
      X           PSP1(IXDIM, IYDIM),      PRP1(IXDIM, IYDIM),                    GI  106
      X           WPOINT(IXDIM, IYDIM, 4), TPOINT(IXDIM, IYDIM, 4),               GI  106
      X           PPOINT(IXDIM, IYDIM, 4), PRPONT(IXDIM, IYDIM, 4),               GI  106
      X           EPOINT(IXDIM, IYDIM, 4)                                         GI  106
        INDERR=0                                                                  GI  109
        DO 110 J = 1, IYDIM, 1                                                    GI  110
        DO 110 I = 1, IXDIM, 1                                                    GI  111
        WPOINT(I,J,1)=.TRUE.                                                      GI  112
        PPOINT(I,J,1)=.TRUE.                                                      GI  113
        PRPONT(I,J,1)=.TRUE.                                                      GI  113
    110 TPOINT(I,J,1)=.TRUE.                                                      GI  114
C        ALL POINTS ARE INITIALLY IN THE FLUID.                                   GI  115
        DO 115 I = 1, IXDIM, 1                                                    GI  120
        EPOINT(I,1,2)=.TRUE.                                                      GI  121
    115 EPOINT(I, IYDIM, 1) = .TRUE.                                              GI  122
        DO 120 I = 1, IYDIM, 1                                                    GI  125
        EPOINT(1, I, 4) = .TRUE.                                                  GI  126
    120 EPOINT(IXDIM, I, 3) = .TRUE.                                              GI  127
C        EDGE POINTS ARE MARKED.                                                  GI  130
        WRITE (6,195)                                                            GI  140
    195 FORMAT ('1','GEOMETRY INPUT BEGUN')                                       GI  141
    200 ICARD=0                                                                   GI  145
        OMEGA=.FALSE.                                                             GI  146
        PSI=.FALSE.                                                               GI  147
        TEMP=.FALSE.                                                              GI  148
        SLIP=.FALSE.                                                              GI  149
        MOVE=.FALSE.                                                              GI  150
        CONT=.FALSE.                                                              GI  151
        PRES=.FALSE.                                                              GI  152
        NAMELIST /SPECS/ ICARD, OMEGA, PSI, TEMP, SLIP, MOVE, CONT, PRES          GI  155
        READ ( 5, SPECS, END = 310, ERR = 202)
        IF (OMEGA) GO TO 205                                                      GI  160
        IF (PSI) GO TO 205                                                        GI  161
        IF (TEMP) GO TO 205                                                       GI  162
        IF (CONT) GO TO 205                                                       GI  163
        IF (PRES) GO TO 205                                                       GI  164
    202 INDERR=1                                                                  GI  165
        WRITE (6,203)                                                            GI  170
    203 FORMAT('0','ERROR IN SPECIFICATION CARD.'/)                               GI  171
        WRITE (6, SPECS)                                                          GI  175
        GO TO 200                                                                 GI  180
    205 IF (ICARD .LE. 0) GO TO 202                                               GI  185
        DO 300 ICOUNT = 1, ICARD, 1                                               GI  195
        ISLOPE = -10                                                              GI  200
        DO 210 L=1,5,1                                                            GI  205
        INDX(L)=0                                                                 GI  210
        INDY(L)=0                                                                 GI  211
    210 INDE(L)=0                                                                 GI  212
        WVAL=0                                                                    GI  213
        PVAL=0                                                                    GI  214
        TVAL=0                                                                    GI  215
        PRVAL = 0                                                                 GI  216
        READ ( 5, 215, END = 310, ERR = 225) ISLOPE, WVAL, PVAL, TVAL,           GI  220
      X    (INDX(L), INDY(L), INDE(L), L = 1, 5, 1), PRVAL                        GI  221
    215 FORMAT (I2, 1X, 3(F6.3, 1X), 5(3I3, 1X), F6.3)                            GI  222
        IPOINT=0                                                                  GI  225
C        IF ISLOPE = -1 THE BOUNDARY SLOPES DOWNWARD.                             GI  228
C        IF ISLOPE =  0 THE BOUNDARY IS HORIZONTAL.                               GI  228
C        IF ISLOPE =  1 THE BOUNDARY SLOPES UPWARD.                               GI  228
C        IF ISLOPE = 10 THE BOUNDARY IS VERTICLE.                                 GI  228
    220 IF (ISLOPE+1)225,233,221                                                  GI  230
    221 IF (ISLOPE)225,232,222                                                    GI  231
    222 IF (ISLOPE-1)225,231,223                                                  GI  232
    223 IF (ISLOPE-10)225,230,225                                                 GI  233
    225 INDERR=1                                                                  GI  235
        WRITE (6,226)                                                            GI  240
    226 FORMAT('0','READ ERROR OR ERROR IN SLOPE SPECIFICATION.',/)              GI  241
        WRITE (6, 227)   ISLOPE, WVAL, PVAL,TVAL,(INDX(L),INDY(L),INDE(L),GI     GI  245
      X               L = 1, 5, 1), PRVAL                                         GI  246
    227 FORMAT (5X, I2, 1X, 3(F6.3, 1X), 5(3I3, 1X), F6.3)                        GI  248
```

```
          GO TO 300                                                    GI  250
      230 IPOINT=IPOINT+1                                              GI  255
      231 IPOINT=IPOINT+1                                              GI  256
      232 IPOINT=IPOINT+1                                              GI  257
      233 IPOINT=IPOINT+1                                              GI  258
C         INDEX FOR COMPUTED GO TO, IPOINT, IS NOW SET UP.            GI  259
C         IF IPOINT = 1, LINE SLOPES DOWNWARD. IF IPOINT = 2, LINE IS HORZ. GI  260
C         IF IPOINT = 3, LINE SLOPES UPWARD.   IF IPOINT = 4, LINE IS VERT.  GI  261
          DO 295 L=1,5,1                                               GI  265
          IUPPER=INDE(L)                                               GI  266
          GO TO (240,240,240,235), IPOINT                             GI  270
      235 ILOW=INDY(L)                                                 GI  275
          GO TO 245                                                    GI  276
      240 ILOW=INDX(L)                                                 GI  280
      245 I=INDX(L)                                                    GI  285
          J=INDY(L)                                                    GI  286
C         INITIAL X&Y COORDINATES NOW ASSIGNED.                       GI  287
          DO 290 INDEX=ILOW,IUPPER,1                                   GI  290
          IF (I) 255,295,250                                           GI  295
      250 IF (J) 255,295,251                                           GI  296
      251 IF (I - IXDIM) 252, 252, 255                                 GI  297
      252 IF (J - IYDIM) 260, 260, 255                                 GI  298
C         IF I OR J = 0 THERE IS NO BOUNDARY SPECIFIED.               GI  298
C         CHECK TO SEE IF THIS POINT IS WITHIN ARRAY BOUNDARIES.      GI  299
      255 INDERR=1                                                     GI  300
          WRITE (6, 256) L, IXDIM, IYDIM                               GI  301
      256 FORMAT ('0ERROR IN COORDINATE SPECIFICATION NUMBER ', I1,   GI  302
         X       '. IXDIM =', I3, ' IYDIM =', I3)                     GI  303
          WRITE (6,227) ISLOPE, WVAL, PVAL, TVAL, (INDX(M), INDY(M),INDE(M), GI  305
         X  M=1,5)                                                     GI  306
          GO TO 295                                                    GI  310
      260 IF (.NOT. OMEGA) GO TO 265                                   GI  315
          WPOINT (I,J,1)=.FALSE.                                       GI  320
          WPOINT (I,J,2)=SLIP                                          GI  321
          WPOINT (I,J,3)=MOVE                                          GI  322
          WP1(I,J)=WVAL                                                GI  323
      265 IF (.NOT. PSI) GO TO 270                                     GI  330
          PPOINT (I,J,1)=.FALSE.                                       GI  335
          PPOINT (I,J,2)=SLIP                                          GI  336
          PPOINT (I,J,3)=MOVE                                          GI  337
          PSP1(I,J)=PVAL                                               GI  338
      270 IF (.NOT. TEMP) GO TO 275                                    GI  345
          TPOINT (I,J,1)=.FALSE.                                       GI  350
          TPOINT (I,J,2)=SLIP                                          GI  351
          TPOINT (I,J,3)=MOVE                                          GI  352
          TP1(I,J)=TVAL                                                GI  353
      275 IF (.NOT. CONT) GO TO 278                                    GI  360
          WPOINT (I,J,4)=.TRUE.                                        GI  365
          PPOINT (I,J,4)=.TRUE.                                        GI  366
          TPOINT (I,J,4)=.TRUE.                                        GI  367
          PRPONT(I, J, 4) = .TRUE.                                     GI  367
      278 IF (.NOT. PRES) GO TO 280                                    GI  368
          PRPONT(I,J,1)=.FALSE.                                        GI  368
          PRPONT(I,J,2)=SLIP                                           GI  368
          PRPONT(I,J,3)=MOVE                                           GI  368
          PRP1(I,J)=PRVAL                                              GI  368
C         ALL INDICATORS & VALUES HAVE NOW BEEN PROPERLY ASSIGNED FOR THE GI  370
C         GIVEN POINT.                                                GI  371
      280 GO TO (286,288,287,285), IPOINT                             GI  375
      285 J=J+1                                                        GI  380
          GO TO 290                                                    GI  381
      286 J=J-1                                                        GI  385
          GO TO 288                                                    GI  386
      287 J=J+1                                                        GI  390
      288 I=I+1                                                        GI  391
      290 CONTINUE                                                     GI  395
C         ALL POINTS FROM ILOW TO INDE(L) ARE NOW ASSIGNED.           GI  396
      295 CONTINUE                                                     GI  400
C         ALL 5 LOCI ON A CARD HAVE NOW BEEN PROCESSED.               GI  401
      300 CONTINUE                                                     GI  405
C         ALL DATA CARDS SPECIFIED BY ICARD ARE NOW PROCESSED.        GI  406
      305 GO TO 200                                                    GI  410
      310 CONTINUE                                                     GI
      350 IF (INDERR .EQ. 0) RETURN 1                                  GI  480
          RETURN 2                                                     GI  485
          END                                                         GI


          SUBROUTINE  PSCALC(PSP1,PS,PSM1,WP1,TP1,PPOINT,EPOINT,
         X       IDIMX,IDIMY ,Q,EPS,MINI,PSP1L,WP1L)
          DIMENSION PSP1L (10), WP1L(10)
          DATA LITWP1/'WP1 '/, LITTP1/'TP1 '/, LIPSP1/'PSP1'/
          REAL PSP1(IDIMX,IDIMY),PS(IDIMX,IDIMY),PSM1(IDIMX,IDIMY),
```

```
      X     WP1(IDIMX,IDIMY),TP1(IDIMX,IDIMY)
      LOGICAL *1 PPOINT(IDIMX,IDIMY,4), EPOINT(IDIMX,IDIMY,4)
      LOGICAL*1 DIRECT
      INTEGER*2 COUNT/0/
      COMMON /INDATA/ A     , F    , DT, DTP, DTPR, UL, UU, D, VIS, CON,
      X               HALG, AM, WN, TEST, DTTS, XLL, YLL, XUR, YUR, XGP,
      X               YGP, IXDIM, IYDIM
      COMMON /USDATA/ AS, FS, CF, AF, ASFS, TM, ITT,R,AL,BT,C,H,TP ,TPR,
      X  CFI
      UNUU = 0
      UNUD = 0
  109 ITT=0
      IF (COUNT .GT. 1 ) GO TO 41
      COUNT = COUNT + 1
      GO TO 104
   41 DO 42 J = 1,IDIMY,1
      DO 42 I = 1,IDIMX,1
   42 PSP1(I,J) = 2.0*PS(I,J) - PSM1(I,J)
  104 DIRECT = .FALSE.
      JDUMMY = IDIMY + 1
  105 IDIFFC=0
      ITT=ITT+1
      IF (ITT .GT.100) GO TO 98
      DO 103 JJ=1,IDIMY,1
   10 DIRECT = .NOT. DIRECT
      IF(      DIRECT) J=JJ
      IF(.NOT. DIRECT) J=JDUMMY-JJ
      DO 100 I=1,IDIMX,1
      IF (PPOINT(I,J,1)) GO TO 1
      IF (PPOINT(I,J,2)) GO TO 100
      IF (PPOINT(I,J,3)) GO TO 200
C     RIGID,NO-SLIP,NOT MOVING   KEEP PSP1 AS GIVEN   ALSO FOR FREE-SLIP
      GO TO 100
    1 IF ( I .EQ. IDIMX) GO TO 2
      IF ( I .EQ. 1    ) GO TO 12
C     NORMAL INTERIOR TO FLUID POINT
      IP = (J-1)*IDIMX+I
      DELP       = (PSP1L(IP+1)+PSP1L(IP-1)+FS*(PSP1L(IP+IDIMX)+PSP1L(IP
      X   -IDIMX) + AS*WP1L(IP)))*CFI
      ABSDIF = ABS(  DELP -PSP1L(IP ))
      PSP1L(IP) =PSP1L(IP) + Q* (DELP - PSP1L(IP))
      IF (ABSDIF .GT.EPS   ) GO TO 5
      IF (ITT .LT. MINI) GO TO 5
      GO TO 100
    5 IDIFFC = 1
      IX = I
      IY = J
      GO TO 100
    2 IF (PPOINT(I,J,4)) GO TO 8
C     PERIODIC   RIGHT
      PSP1(I,J) = PSP1(I-(IDIMX-1),J)
      GO TO 100
C     CONTINUATIVE   RIGHT
    8 PSP1(I,J) = (PSP1(I,J+1)+PSP1(I,J-1)+AS*WP1(I,J))*0.5
      GO TO 100
   12 IF (PPOINT(I,J,4)) GO TO 18
C     PERIODIC  LEFT
      DELP       = (PSP1(I+1,J)+PSP1(I-1+(IDIMX-1),J)+FS*(PSP1(I,J+1)+
      X             PSP1(I,J-1)+AS*WP1(I,J)))*CFI
      PSP1(I,J) =PSP1(I,J) + Q* (DELP - PSP1(I,J))
      GO TO 100
C     CONTINUATIVE  LEFT
C  18 PSP1(I,J) = (PSP1(I+2,J) + PSP1(I,J) + FS * (PSP1(I+1,J+1)
C     X            + PSP1(I+1,J-1) + AS * WP1(I+1,J))* CFI
   18 PSP1(I,J) = (PSP1(I   ,J) + PSP1(I,J) + FS * (PSP1(I   ,J+1)
      X            + PSP1(I,  J-1) + AS * WP1(I   ,J)))* CFI
      GO TO 100
  200 IF ( J .EQ. IDIMY) UNUU=1
      IF ( J .EQ. 1    ) UNUD=1
  100 CONTINUE
      IF(      DIRECT) GO TO 10
  103 CONTINUE
      IF (UNUD .EQ. 0) GO TO 101
      PSS = 0.0
      DO 205 I=1,IDIMX,1
  205 PSS = PSS + PSP1(I,  2  ) + 0.5*AS*WP1(I,1)
      PSS = (PSS/IDIMX) -A*UL
      DO 207 I=1,IDIMX,1
  207 PSP1(I,  1  ) = PSS
  101 IF (UNUU .EQ. 0) GO TO 102
      PSS = 0.0
      DO 209 I=1,IDIMX,1
  209 PSS = PSS + PSP1(I,IDIMY-1) + 0.5*AS*WP1(I,IDIMY)
      PSS = (PSS/IDIMX) +A*UU
```

```
      DO 211 I=1,IDIMX,1
211 PSP1(I,IDIMY) = PSS
      IY =(IDIMY/2) +1
      TEMPP = PSP1(1,IY)
      DO 219 J=1,IDIMY,1
      DO 219 I=1,IDIMX,1
219  PSP1(I,J) = PSP1(I,J) - TEMPP
102 IF (IDIFFC.GT.0) GO TO 105
      RETURN
 98 WRITE (6,99)IX,IY
 99 FORMAT ('1','THIS HAS TAKEN OVER 200 ITS., CHECK POINT ',I2,1X,I2,
    X'  PSI SOL IS INCOMPLETE')
      CALL DATAPR (PSP1,LIPSP1,IDIMX,IDIMY)
      CALL DATAPR (WP1 ,LITWP1,IDIMX,IDIMY)
      RETURN
      END



      SUBROUTINE WF(WP1,PSP1,WPOINT,EPOINT,IDIMX,IDIMY)
      REAL WP1(IDIMX,IDIMY),                    PSP1(IDIMX,IDIMY)
      LOGICAL *1 WPOINT(IDIMX,IDIMY,4), EPOINT(IDIMX,IDIMY,4)
      COMMON /INDATA/ A    , F    , DT, DTP, DTPR, UL, UU, D, VIS, CON,
    X              HALG, AM, WN, TEST, DTTS, XLL, YLL, XUR, YUR, XGP,
    X              YGP, IXDIM, IYDIM,TDIF, APG
      COMMON /USDATA/ AS, FS, CF, AF, ASFS, TM, ITT,R,AL,BT,C,H,TP ,TPR,
    X CFI
C        NO SLIP SURFACES ARE UPDATED HERE
      DO 100 J=1,IDIMY,1
      DO 100 I=1,IDIMX,1
      IF (WPOINT(I,J,1))GO TO 100
      IF (WPOINT(I,J,2))GO TO 50
      IF (WPOINT(I,J,3))GO TO 200
      IF ( J .EQ. IDIMY) GO TO 10
      IF ( J .EQ. 1    ) GO TO 20
      IF ( I .EQ. IDIMX) GO TO 30
      IF ( I .EQ. 1  ) GO TO 40
C        THESE ARE INTERIOR NO SLIP SURFACES
      IF (.NOT. WPOINT(I+1,J,1)) GO TO 1
      WA =(PSP1(I,J)-PSP1(I+1,J))/ASFS
 13 IF (.NOT. WPOINT(I-1,J,1)) GO TO 3
      WB =(PSP1(I,J)-PSP1(I-1,J))/ASFS
 12 IF (.NOT. WPOINT(I,J+1,1)) GO TO 2
      WC =(PSP1(I,J)-PSP1(I,J+1))/AS
 14 IF (.NOT. WPOINT(I,J-1,1)) GO TO 4
      WD =(PSP1(I,J)-PSP1(I,J-1))/AS
110 WP1(I,J) = WA+WB+WC+WD
      GO TO 100
  1 WA =(PSP1(I,J)-PSP1(I-1,J))/ASFS
      GO TO 13
  3 WB =(PSP1(I,J)-PSP1(I+1,J))/ASFS
      GO TO 12
  2 WC =(PSP1(I,J)-PSP1(I,J-1))/AS
      GO TO 14
  4 WD =(PSP1(I,J)-PSP1(I,J+1))/AS
      GO TO 110
C        THESE ARE NO SLIP SURFACES AT EDGES
 10 WP1(I,J) = 2.0*(PSP1(I,J)-PSP1(I,J-1))/AS
      GO TO 100
 20 WP1(I,J) = 2.0*(PSP1(I,J)-PSP1(I,J+1))/AS
      GO TO 100
 50 WP1(I,J) = 0.0
      GO TO 100
 30 WA =(PSP1(I,J)-PSP1(I-1,J))/ASFS
      WB = WA
      GO TO 12
 40 WA =(PSP1(I,J)-PSP1(I+1,J))/ASFS
      WB = WA
      GO TO 12
200 IF ( J .EQ. IDIMY)  WP1(I,J) = 2.0*((PSP1(I,J)-PSP1(I,J-1))/A-UU)
    X     /A
      IF ( J .EQ. 1    )  WP1(I,J) = 2.0*(UL-(PSP1(I,J+1)-PSP1(I,J))/A)
    X     /A
100 CONTINUE
      RETURN
      END



      SUBROUTINE STAB(PSP1, EPOINT, IDIMX, IDIMY,PS ,PSM1,*,*)
      REAL PSP1(IDIMX,IDIMY)             ,
    X      PS  (IDIMX,IDIMY)             ,
    X      PSM1(IDIMX,IDIMY)
      LOGICAL *1 EPOINT(IDIMX,IDIMY,4)
```

```
      COMMON /INDATA/ A    , F    , DT, DTP, DTPR, UL, UU, D, VIS, CON,
     X                HALG, AM, WN, TEST, DTTS, XLL, YLL, XUR, YUR, XGP,
     X                YGP, IXDIM, IYDIM, TDIF, APG
      COMMON /USDATA/ AS, FS, CF, AF, ASFS, TM, ITT,R,AL,BT,C,H,TP ,TPR,
     X CFI
      SMAX = 0.0
      DO 100 J=1,IDIMY,1
      DO 100 I=1,IDIMX,1
      IF (EPOINT(I,J,2)) GO TO 100
      UST = ABS(PSP1(I,J) - PSP1(I,J-1))
      IF (UST .GT. SMAX)  SMAX = UST
  100 CONTINUE
      DO 200 J=1,IDIMY,1
      DO 200 I=1,IDIMX,1
      IF (EPOINT(I,J,4)) GO TO 200
      UST = ABS(PSP1(I,J) - PSP1(I-1,J))
      IF (UST .GT. SMAX)  SMAX = UST
  200 CONTINUE
    9 IF ((DT*SMAX/(AF*A))-.50) 10,10,20
   10 TM = TM + DT
      RETURN 1
   20 DT=DT/2.0
      C = VIS*DT/AS
      H = CON*DT/AS
      AL = DT/AF
      BT = DT/A
      GO TO 9
      END


      SUBROUTINE WUCALC(WP1,W,WM1,TP1,PSP1,PSM1,SS,S3,WPOINT,EPOINT,    WU04
     X         IDIMX,IDIMY)
      REAL WP1(IDIMX,IDIMY), W(IDIMX,IDIMY), PSP1(IDIMX,IDIMY),
     X     TP1(IDIMX,IDIMY), PSM1(IDIMX,IDIMY),SS(IDIMX,IDIMY),
     X                       WM1(IDIMX,IDIMY),S3(IDIMX,IDIMY)
      LOGICAL *1 WPOINT(IDIMX,IDIMY,4) ,EPOINT(IDIMX,IDIMY,4)
      COMMON /INDATA/ A    , F    , DT, DTP, DTPR, UL, UU, D, VIS, CON,
     X                HALG, AM, WN, TEST, DTTS, XLL, YLL, XUR, YUR, XGP,
     X                YGP, IXDIM, IYDIM
      COMMON /USDATA/ AS, FS, CF, AF, ASFS, TM, ITT,R,AL,BT,C,H,TP ,TPR,
     X CFI
      DATA A11/0.583333333/ ,  A12/0.625/,
     X     A13/-.083333333/, A14/-.125/,
     X     A21/-.083333333/, A22/-.0416666667/ ,
     X     A23/0.083333333/ , A24/0.0416666667/
C     A11=7/12,A12=5/8,A13=1/2-A11,A14=1/2-A12,A21=A13,A22=A14/3,A23=-A21
C     A24=-A22
      DO 301 J=1,IDIMY,1
      DO 301 I=1,IDIMX,1
      IF ( I .EQ. IDIMX) GO TO 301
      UF = (PSP1(I,J+1) - PSP1(I,J-1)+PSP1(I+1,J+1)-PSP1(I+1,J-1))
     X         /(4.0*A)
      IF ( J .EQ. IDIMY) UF=UU
      IF ( J .EQ. 1    ) UF=UL
      ALF = UF*AL
      IF (ALF .LT. 0.0) GO TO 311
      IF (.NOT. WPOINT(I,J,1)) GO TO 600
      IF (        WPOINT(I  ,J,4)) GO TO 600
      IF (        WPOINT(I+1,J,4)) GO TO 600
      IF ( I .EQ. 1    ) GO TO 400
      IF (.NOT. WPOINT(I-1,J,1))GO TO 200
      IF (        WPOINT(I-1,J,4))GO TO 200
      IF (.NOT. WPOINT(I+1,J,1))GO TO 200
      IF (        WPOINT(I+1,J,4))GO TO 200
  400 AA=ALF*ALF
      AAA=ALF*AA
      AAAA = ALF*AAA
      A1=A11*ALF
      A2=A12*AA
      A3=A13*AAA
      A4=A14*AAAA
      B1=A21*ALF
      B2=A22*AA
      B3=A23*AAA
      B4=A24*AAAA
      B =ALF-1.0
      BB = B*B
      BBB = B*BB
      BBBB = B*BBB
      C1= A11*B + A11
      C2= A12*BB - A12
      C3= A13*BBB + A13
      C4= A14*BBBB - A14
```

```
        D1= A21*B
        D2= A22*BB
        D3= A23*BBB
        D4= A24*BBBB
        IF ( I  .EQ. 1    ) GO TO 304
        IF ( I-1.EQ. 1    ) GO TO 305
        IF ( I+1.EQ.IDIMX) GO TO 309
        FDP= WM1(I-2,J) + WM1(I+1,J)
        FDM= WM1(I-2,J) - WM1(I+1,J)
308     FBP= WM1(I-1,J) + WM1(I+2,J)
        FBM= WM1(I-1,J) - WM1(I+2,J)
307     FCP= WM1(I-1,J) + WM1(I  ,J)
        FCM= WM1(I-1,J) - WM1(I  ,J)
306     FAP= WM1(I  ,J) + WM1(I+1,J)
        FAM= WM1(I  ,J) - WM1(I+1,J)
        TF = A1*FAP + A2*FAM + A3*FAP + A4*FAM
  X        + B1*FBP + B2*FBM + B3*FBP + B4*FBM
  X        + C1*FCP + C2*FCM + C3*FCP + C4*FCM
  X        + D1*FDP + D2*FDM + D3*FDP + D4*FDM
        WP1(I,J) = WP1(I,J) - 0.5*TF
        WP1(I+1,J) = WP1(I+1,J) + 0.5*TF
        IF ( I+1.LT.IDIMX) GO TO 302
        IF (.NOT. WPOINT(I+1,J,1)) GO TO 301
C       PERIODIC RIGHT
        WP1(I+1-(IDIMX-1),J) = WP1(I+1-(IDIMX-1),J) + 0.5*TF
        GO TO 301
302     IF ( I  .GT. 1    ) GO TO 301
        IF (.NOT. WPOINT(I  ,J,1)) GO TO 301
C       PERIODIC LEFT
        WP1(I  +(IDIMX-1),J) = WP1(I  +(IDIMX-1),J) - 0.5*TF
        GO TO 301
304     FDP = WM1(I-2+(IDIMX-1),J) + WM1(I+1,J)
        FDM = WM1(I-2+(IDIMX-1),J) - WM1(I+1,J)
        FBP = WM1(I-1+(IDIMX-1),J) + WM1(I+2,J)
        FBM = WM1(I-1+(IDIMX-1),J) - WM1(I+2,J)
        FCP = WM1(I-1+(IDIMX-1),J) + WM1(I  ,J)
        FCM = WM1(I-1+(IDIMX-1),J) - WM1(I  ,J)
        GO TO 306
305     FDP = WM1(I-2+(IDIMX-1),J) + WM1(I+1,J)
        FDM = WM1(I-2+(IDIMX-1),J) - WM1(I+1,J)
        GO TO 308
309     FBP = WM1(I-1,J) + WM1(I+2-(IDIMX-1),J)
        FBM = WM1(I-1,J) - WM1(I+2-(IDIMX-1),J)
        FDP = WM1(I-2,J) + WM1(I+1,J)
        FDM = WM1(I-2,J) - WM1(I+1,J)
        GO TO 307
311     IF (.NOT. WPOINT(I+1,J,1)) GO TO 601
        IF (      WPOINT(I+1,J,4)) GO TO 601
        IF (      WPOINT(I  ,J,4)) GO TO 601
        K = I+1
        IF ( K .EQ. IDIMX) GO TO 401
        IF (.NOT. WPOINT(K+1,J,1)) GO TO 201
        IF (      WPOINT(K+1,J,4)) GO TO 201
        IF (.NOT. WPOINT(K-1,J,1)) GO TO 201
        IF (      WPOINT(K-1,J,4)) GO TO 201
401     AA=ALF*ALF
        AAA=ALF*AA
        AAAA = ALF*AAA
        A1=A11*ALF
        A2=A12*AA
        A3=A13*AAA
        A4=A14*AAAA
        B1=A21*ALF
        B2=A22*AA
        B3=A23*AAA
        B4=A24*AAAA
        B=ALF+1.0
        BB = B*B
        BBB = B*BB
        BBBB = B*BBB
        C1= A11*B - A11
        C2= A12*BB - A12
        C3= A13*BBB - A13
        C4= A14*BBBB - A14
        D1= A21*B
        D2= A22*BB
        D3= A23*BBB
        D4= A24*BBBB
        IF ( K .EQ. IDIMX) GO TO 314
        IF ( K+1.EQ.IDIMX) GO TO 315
        IF ( K-1.EQ. 1    ) GO TO 319
        FDP = WM1(K-1,J) + WM1(K+2,J)
        FDM = WM1(K-1,J) - WM1(K+2,J)
318     FBP = WM1(K-2,J) + WM1(K+1,J)
```

```
      FBM = WM1(K-2,J) - WM1(K+1,J)
  317 FCP = WM1(K  ,J) + WM1(K+1,J)
      FCM = WM1(K  ,J) - WM1(K+1,J)
  316 FAP = WM1(K-1,J) + WM1(K  ,J)
      FAM = WM1(K-1,J) - WM1(K  ,J)
      TF = A1*FAP + A2*FAM + A3*FAP + A4*FAM
  X    + B1*FBP + B2*FBM + B3*FBP + B4*FBM
  X    + C1*FCP + C2*FCM + C3*FCP + C4*FCM
  X    + D1*FDP + D2*FDM + D3*FDP + D4*FDM
      WP1(K-1,J) = WP1(K-1,J) - 0.5*TF
      WP1(K  ,J) = WP1(K  ,J) + 0.5* TF
      IF ( K .LT. IDIMX) GO TO 312
      IF (.NOT. WPOINT(K  ,J,1)) GO TO 301
C     PERIODIC RIGHT
      WP1(K  -(IDIMX-1),J) = WP1(K  -(IDIMX-1),J) + 0.5*TF
      GO TO 301
  312 IF ( K-1.GT. 1   ) GO TO 301
      IF (.NOT. WPOINT(K-1,J,1)) GO TO 301
C     PERIODIC LEFT
      WP1(K-1+(IDIMX-1),J) = WP1(K-1+(IDIMX-1),J) - 0.5*TF
      GO TO 301
  314 FDP = WM1(K-1,J) + WM1(K+2-(IDIMX-1),J)
      FDM = WM1(K-1,J) - WM1(K+2-(IDIMX-1),J)
      FBP = WM1(K-2,J) + WM1(K+1-(IDIMX-1),J)
      FBM = WM1(K-2,J) - WM1(K+1-(IDIMX-1),J)
      FCP = WM1(K  ,J) + WM1(K+1-(IDIMX-1),J)
      FCM = WM1(K  ,J) - WM1(K+1-(IDIMX-1),J)
      GO TO 316
  315 FDP = WM1(K-1,J) + WM1(K+2-(IDIMX-1),J)
      FDM = WM1(K-1,J) - WM1(K+2-(IDIMX-1),J)
      GO TO 318
  319 FBP = WM1(K-2+(IDIMX-1),J) + WM1(K+1,J)
      FBM = WM1(K-2+(IDIMX-1),J) - WM1(K+1,J)
      FDP = WM1(K-1,J) + WM1(K+2,J)
      FDM = WM1(K-1,J) - WM1(K+2,J)
      GO TO 317
  600 TF = ALF * WM1(I,J)
      GO TO 602
  601 TF = ALF * WM1(I+1,J)
  602 WP1(I,J) = WP1(I,J) - TF
      WP1(I+1,J) = WP1(I+1,J) + TF
      IF (WPOINT(I,J,4)) WP1(I,J) = WP1(I,J) + TF
      IF(WPOINT(I+1,J,4))WP1(I+1,J)=WM1(I+1,J)+ALF*(WM1(I,J)-WM1(I+1,J))
      GO TO 301
  200 A1 = 0.25*ALF
      A2 = A1 * ALF
      B2   = A2 - 2.0*A1
      TRP = WM1(I-1,J) + WM1(I,J)
      TRM = WM1(I-1,J) - WM1(I,J)
      TPL = WM1(I  ,J) + WM1(I+1,J)
      TML = WM1(I  ,J) - WM1(I+1,J)
      TF  = A1 *TPL +A1 *TRP + A2 *TML + B2  *TRM
      WP1(I+1,J) = WP1(I+1,J) + TF
      WP1( I ,J) = WP1(I  ,J) - TF
      GO TO 301
  201 A1 = 0.25*ALF
      A2 = A1 * ALF
      B2   = A2  + 2.0*A1
      TRP = WM1(K  ,J) + WM1(K+1,J)
      TRM = WM1(K  ,J) - WM1(K+1,J)
      TPL = WM1(K-1,J) + WM1(K  ,J)
      TML = WM1(K-1,J) - WM1(K  ,J)
      TF  = A1 *TPL +A1 *TRP + A2 *TML + B2  *TRM
      WP1(K  ,J) = WP1(K  ,J) + TF
      WP1(K-1,J) = WP1(K-1,J) - TF
  301 CONTINUE
      RETURN
      END


      SUBROUTINE WDCALC(WP1,W,WM1,TP1,PSP1,PSM1,SS,S3,WPOINT,EPOINT,
  X              IDIMX,IDIMY)
      REAL WP1(IDIMX,IDIMY), W(IDIMX,IDIMY), PSP1(IDIMX,IDIMY),
  X      TP1(IDIMX,IDIMY), PSM1(IDIMX,IDIMY),SS(IDIMX,IDIMY),
  X              WM1(IDIMX,IDIMY),S3(IDIMX,IDIMY)
      LOGICAL *1 WPOINT(IDIMX,IDIMY,4), EPOINT(IDIMX,IDIMY,4)
      COMMON /INDATA/ A    , F    , DT, DTP, DTPR, UL, UU, D, VIS, CON,
  X              HALG, AM, WN, TEST, DTTS, XLL, YLL, XUR, YUR, XGP,
  X              YGP, IXDIM, IYDIM,TDIF, APG
      COMMON /USDATA/ AS, FS, CF, AF, ASFS, TM, ITT,R,AL,BT,C,H,TP ,TPR,
  X  CFI
      DO  90 J=1,IDIMY,1
      DO  90 I=1,IDIMX,1
```

```
      IF (WPOINT(I,J,1)) GO TO 90
      WP1(I,J) = W  (I,J)
   90 CONTINUE
      DO 100 J=1,IDIMY,1
      DO 100 I=1,IDIMX,1
      IF (WPOINT(I,J,1)) GO TO 1
      IF (WPOINT(I,J,2)) GO TO 101
      IF (WPOINT(I,J,3)) GO TO 100
C     RIGID,NO-SLIP,NOT MOVING
      GO TO 101
    1 IF (EPOINT(I,J,3)) GO TO 2
      IF (EPOINT(I,J,4)) GO TO 12
      IF (EPOINT(I,J,1)) GO TO 200
      IF (EPOINT(I,J,2)) GO TO 200
C     NORMAL INTERIOR TO FLUID POINT
      WP1(I,J) = WP1(I,J)+C*  (W(I,J+1)+  W(I,J-1)+  (W(I+1,J)
     X          + W(I-1,J)- CF*  W(I,J))/FS)
      GO TO 100
    2 IF (WPOINT(I,J,4)) GO TO 8
C     PERIODIC RIGHT
      WP1(I,J) = WP1(I,J)+ C* (W(I,J+1)+  W(I,J-1)+  (W(I+1-(IDIMX-1),J)
     X          + W(I-1,J)- CF* W(I,J))/FS)
      GO TO 100
C     CONTINUATIVE RIGHT
C   8 WP1(I,J) = (DT/(2.0*A*AF))* ((PSP1(I,J+1)-PSP1(I,J-1))*(W(I-1,J)
C     X          -(W(I,J+1)+W(I,J-1))*0.5) + (PSP1(I-1,J)-PSP1(I,J))*
C     X           (W(I,J-1)-W(I,J+1)))+0.5*(W(I,J+1)+W(I,J-1))
    8 WP1(I,J) = WP1(I  ,J) + C * (W(I  ,J+1) + W(I ,J-1)-2.0*W(I,J))
      GO TO 100
   12 IF (WPOINT(I,J,4))GO TO 18
C     PERIODIC LEFT
      WP1(I,J) = WP1(I,J)+ C* (W(I,J+1)+ W(I,J-1)+  (W(I+1,J)
     X          + W(I-1+(IDIMX-1),J) - CF* W(I,J))/FS)
      GO TO 100
C     CONTINUATIVE LEFT
C  18 WP1(I,J) = WP1(I+1,J) + C * (W(I+1,J+1) + W(I+1,J-1)
C     X          + (W(I+2,J) + W(I,J) - CF * W(I+1,J))/FS)
   18 WP1(I,J) = WP1(I  ,J) + C * (W(I  ,J+1) + W(I ,J-1)
     X          + (W(I  ,J) + W(I,J) - CF * W(I ,J))/FS)
      GO TO 100
C     FREE SLIP - NO DIFFUSION
  101 GO TO 100
  100 CONTINUE
      IF (HALG .EQ. 0.0) GO TO 51
      DO 50  J=1,IDIMY,1
      DO 50  I=1,IDIMX,1
      IF (EPOINT(I,J,3))GO TO 52
      IF (EPOINT(I,J,4))GO TO 62
      WP1(I,J) = WP1(I,J) + HALG * AL * (TP1(I+1,J)-TP1(I-1,J))
      GO TO 50
   52 IF (WPOINT(I,J,4)) GO TO 50
C     PERIODIC RIGHT
      WP1(I,J) = WP1(I,J) + HALG * AL*(TP1(I+1-(IDIMX-1),J)-TP1(I-1,J))
      GO TO 50
   62 IF (WPOINT(I,J,4)) GO TO 50
C     PERIODIC LEFT
      WP1(I,J) = WP1(I,J) + HALG * AL*(TP1(I+1,J)-TP1(I-1+(IDIMX-1),J))
   50 CONTINUE
   51 RETURN
  200 WRITE (6,201)
  201 FORMAT ('0','CODE FOR MOVING WALL NOT YET WRITTEN')
      RETURN
      END
```

Chapter 3

Computer Simulation of Diffusion Problems Using the Continuous System Modeling Program Language

Farid F. Abraham

IBM Scientific Center
Palo Alto, California

and

Consulting Associate Professor
Materials Science Department
Stanford University
Stanford, California

I. INTRODUCTION

We present an introduction to System 360 Continuous System Modeling Program (S/360-CSMP), a digital simulation language oriented toward the solution of differential equations. The intent is to acquaint the reader with the simplicity of powerful computer software that is at the disposal of the researcher. Designed for use specifically by the scientist or engineer, S/360-CSMP requires only a minimum knowledge of computer programming and operation. Only an elementary knowledge of FORTRAN is necessary. Two simple diffusion problems are discussed to illustrate the basic approach to the simulation method of analysis and the power and ease of computer simulation using a digital simulation language such as S/360-CSMP.

The basic approach presented in this chapter has been expanded into a quarter (ten weeks) course in the Materials Science Department at Stanford University. The computer simulation method of problem solving is applied to a variety of typical problems in engineering. It is assumed that the student has a familiarity with the fundamentals of heat and mass transfer, chemical kinetics, hydrodynamics, and thermodynamics.

The format for the course is the following:

1. Discuss a class of complex physical problems that may be expressed in a mathematical description by a coupled set of ordinary differential equations, a coupled set of partial differential equations, or some combination of these types.
2. Use S/360 CSMP as an appropriate simulation language for the solution of these types of problems. The instructor and students are not burdened by a need for mastering numerical analysis and for writing complex FORTRAN computer programs. We may go directly to the discussion of computer modeling. (Of course, some numerical analysis must be presented.)
3. The course is problem oriented, with the underlying goal that the student can learn to simulate complex applied science problems.

An outline for the Stanford Materials Science course is presented:

1. Introduction to modeling of complex physical systems.
2. S/360 Continuous System Modeling Program language.
3. Simple problems to develop the student's confidence in S/360-CSMP.
   A. Two first-order irreversible chemical reactions.
   B. The penbob (pendulum on a spring).
   C. The Kepler orbit problem.
4. Partial differential equations and finite differencing—general discussion.
5. The vibrating string—discussion of wavelength cut-off and dispersion due to finite differencing.
6. The multistate kinetics of homogeneous nucleation.
7. Transport processes—general discussion.
8. Heat transfer in an insulated bar.
9. Stability analysis of the difference equations. Example: the 1-D heat diffusion equation.
10. Unsteady-state heat conduction in a solidifying alloy.
11. Finite-difference approximation at the stationary interface between two different media. Example: the thermal stabilization of austenite.

12. The freezing of a liquid; the dynamics of the freezing interface.
13. The dynamics of the freezing interface described as a nonlinear heat diffusion problem. Example: 1-D ice formation with particular reference to periodic growth and decay.
14. Ledge dynamics in crystal evaporation.
15. Flat-plate single current heat exchanger.
16. Countercurrent heat exchanger.
17. Two-dimensional heat diffusion.
18. Solute redistribution on freezing.
19. Heat conduction across an irregular boundary.
20. Grain boundary diffusion.

After the discussion of a particular sample problem, the student performs model changes relating to physical processes not considered in class. In this way the student "experiments" with modeling and studies the behavior of the system through its coupling with various subsystems. For example, in the flat-plate single current heat exchanger with laminar flow, we may consider the fluid as a gas containing reactants A and B. We assume that the reaction

$$A + B \xrightarrow{k_1} C + D$$

takes place, where k_1 is a function of temperature, and that the reaction is highly exothermic. We may study the difference in the temperature evolution of the system with and without the reaction and for different k_1.

During the latter half of the quarter, the student works on a "project" problem of his choice, which is, hopefully, related to his research interests. Some examples of past project problems are the following:

1. The variation of the concentration and temperature profiles during the initial transient period of the GaAs crystal growth from solution.
2. The heating characteristics of an externally heated cold seal pressure vessel.
3. A study of the thermal accommodation coefficient of an atom impinging onto a stationary semi-infinite chain of atoms with harmonic potential interactions.

It is the purpose of the course to give the student a powerful tool to facilitate his engineering research activities during his Ph.D. program and after graduation.

II. SYSTEM/360 CONTINUOUS SYSTEM MODELING PROGRAM (S/360 CSMP)*

The System/360 Continuous System Modeling Program (S/360 CSMP) enables a user to solve a physical problem in which the physical system has a differential equation representation. The program provides a basic set of functions, and it also accepts FORTRAN statements. A translator converts the S/360 CSMP statements into a FORTRAN subroutine, which is then compiled and executed alternately with a selected integration routine to solve the differential equations.

Designed specifically for the engineer or scientist, it requires only a minimum knowledge of computer programming and operation. The program provides a basic set of 34 functions. This set is augmented by the FORTRAN library functions (e.g., sine and cosine). Special functions can be defined either through FORTRAN or through a macro capability.

Input and output are simplified by means of a free format for data entry and user-oriented input and output control statements. Output options include printing of variables in a tabular format, print-plotting in graphic form, and preparation of a data set for plotting programs supplied by the user.

Two important features of S/360 CSMP are statement sequencing and a choice of integration methods. With few exceptions, structure statements may be written in any order and may be sorted automatically by the system (at the user's option) in order to establish the proper calculational sequence.

Another feature is the ability to initialize variables or parameters, i.e., to specify that a group of structure statements is to be executed only once at the start of the simulation. All calculations are done in single precision, floating point arithmetic unless otherwise specified. We will present in this section a brief discussion of the S/360 CSMP language.

A. Types of Statements

We define three types of basic statements used in this language:

1. Structure Statements—These describe the functional relationships between the variables of the model.

---

* Based on program description: *System/360 Continuous System Modeling Program User's Manual* (H20-0367-2), IBM Corporation, Data Processing Division, 112 E. Post Road, White Plains, New York, 1966. Excepts from this *User's Manual* © 1967 and 1968, are reprinted by permission of International Business Machines Corporation.

2. Data Statements—These assign numerical values to the parameters, initial conditions, and table entries associated with the problem.
3. Control Statements—These specify options relating to the translation, execution, and output phases of the S/360 CSMP program, such as run time, integration interval, and output variables to be printed.

B. Elements of a Statement

Five elements constitute these S/360 CSMP language statements:

1. Constants—These are unchanging quantities used in their numerical form in the source statements.
2. Variable Names (Also Called "Variables")—These are symbolic representations of quantities that may either change during a run or be changed under program control, between successive runs of the same model structure. Some examples are DIST, RATE2, and MASS.
3. Operators—These are used to indicate the basic arithmetic functions or relationships. As in FORTRAN, these operators are

Symbol	Function
+	Addition
−	Subtraction
*	Multiplication
/	Division
**	Exponentiation
( )	Grouping of variables and/or constants
=	Replacement

Some illustrations of the use of operators to construct structure statements are

$$DIST = RATE * TIME$$
$$Y = A * X ** 2 + B$$
$$A = (B * C) + (D * E)$$

4. Functional Blocks (Functions)—These perform more complex mathematical operations such as integration, time delay, etc. An example of their use is

$$Y = INTGRL(10.0, X)$$

which states that the variable Y is equal to the integral of the variable X, with the initial condition that Y at time zero is equal to the constant quantity 10.0.

5. Labels—The first word of S/360 CSMP data and control statements is a label that identifies to the program the purpose of the statement. For example, to specify the integration interval and the "finish time" for a run, we would use the TIMER statement as follows:

$$\text{TIMER}\quad \text{DELT} = 0.025,\ \text{FINTIM} = 450.0$$

C. Important Features of S/360 CSMP

Three important features of S/360 CSMP are statement sequencing, a choice of integration methods, and freedom from formatting details.

1. With few exceptions, structure statements may be written *in any order* and may be automatically sorted by the system to establish the correct calculation flow.

2. A choice of integration methods may be made between the fifth-order Milne predictor-corrector, fourth-order Runge-Kutta, Simpson's, second-order Adam's, trapezoidal, and rectangular integration methods. The first two methods allow the integration interval to be adjusted by the system to meet a specified error criterion.

3. Input is simplified by means of a free format for data entry and input and output control statements. Data and control statements may be entered in any order and may be intermixed with structure statements. A fixed format for data output at selected increments of the independent variable is provided for all output options. Output options include printing of variables in tabular format, print plotting in graphic form, and preparation of a data set for user-prepared plotting programs.

D. The S/360 CSMP Library of Functions

Table I* contains a few of the functions available in S/360-CSMP. In addition, all the functions available in the standard FORTRAN IV library in the user's system can be treated as functions in the CSMP complement. Illustrations of the most useful functions of FORTRAN are shown in Table II* in the functional notation. The functions INTGRL and IMPL

---

* Courtesy of International Business Machines Corporation.

Table I. Library of S/360 CSMP Functions

Mathematical functions	
General form	Function
Y = INTGRL (IC, X) Y(0) = IC Integrator	$Y = \int_0^t X\,dt + \text{IC}$ Equivalent Laplace transform: $1/S$
Y = DERIV (IC, X) $\dot{X}$(t = 0) = IC Derivative	$Y = dX/dt$ Equivalent Laplace transform: S
Y = DELAY (N, P, X) P = delay time N = number of points sampled in interval P (integer constant) Dead time (delay)	$Y(t) = X(t - P)$　　$t \geq P$ $Y\ \ \ = 0$　　　　$t < P$ Equivalent Laplace transform: e^{-PS}
Y = ZHOLD (X_1, X_2) Zero-order hold	$Y\ \ \ = X_2$　　　$X_1 > 0$ $Y\ \ \ = \text{last output}$　$X_1 \leq 0$ $Y(0) = 0$ Equivalent Laplace transform: $(1/S)(1 - e^{St})$
Y = IMPL (IC, P, FOFY) IC = first guess P = error bound FOFY = output name of last statement in algebraic loop definition Implicit function	$Y = \text{funct }(Y)$ $\|Y - \text{funct }(Y)\| \leq P\|Y\|$

Table II. FORTRAN Functions

General form	Function
Y = EXP (X) Exponential	$Y = e^X$
Y = ALOG (X) Natural logorithm	$Y = \ln(X)$
Y = ALOGIO (X) Common logorithm	$Y = \log_{10}(X)$
Y = ATAN (X) Arctangent	$Y = \arctan(X)$
Y = SIN (X) Trigonometric sine	$Y = \sin(X)$
Y = COS (X) Trigonometric cosine	$Y = \cos(X)$
Y = SQRT (X) Square root	$Y = X^{1/2}$
Y = TANH (X) Hyperbolic tangent	$Y = \tanh(X)$
Y = ABS (X) Absolute value (real argument and output)	$Y = \|X\|$
Y = IABS (X) Absolute value (integer argument and output)	$Y = \|X\|$

of Table I deserve special note for their importance in solving typical simulation problems.

E. The S/360 CSMP Library of Data and Control Statements

Generally, the user would logically first prepare the structure statements and follow these by the data and control statements, in that order. Data statements can be used to assign numeric values to those variables that are to be fixed during a given run. The advantage of assigning variable names and using data statements to specify numeric values is that the latter can be changed, automatically, between successive runs of the same model structure. An example of a data statement is

$$PARAMETER \quad RATE = 550.0, \quad DIST = 1000.0$$

where PARAMETER is the label identifying the card as a parameter card, RATE and DIST are the variables to be assigned numeric values, and 550.0 and 1000.0 are, respectively, the values assigned. Different types of data can be specified by the following labels:

Label	Type of Data
PARAMETER	Parameters
CONSTANT	Constants
INCON	Initial conditions
TABLE	Entries in a stored array

Finally, the user prepares control statements to specify certain operations associated with the translation, execution, and output phases of the program. Control statements, like data statements, can be changed between runs under control of the simulation program. An example of a control statement is

$$PRINT \quad X, XDOT, X2DOT$$

where PRINT is a card label specifying that lists of the variables X, XDOT, and X2DOT are to be printed. Examples of the other control labels are

Label	Control Operation
NOSORT	Specify areas of the structure statements that are not to be sorted
SORT	
INITIAL	Define a block of computations that is to be executed only at the beginning of the run

Label	Control Operation
DYNAMIC	Computations are made repeatedly for each time step
TIMER	Specify the print interval, print-plot interval, finish time, and integration interval
FINISH	Specify a condition for termination of the run
RELERR	Specify relative error for the integration routine
ABSERR	Specify absolute error for the integration routine
METHOD	Specify the integration method
PRINT	Identify variables to be printed
PRTPLT	Identify variables to be print-plotted
TITLE	Print page headings for printed output
LABEL	Print page headings for print-plot output
RANGE	Obtain minimum and maximum values of specified variables

F. Integration Methods

S/360 CSMP uses centralized integration to ensure that all integrator outputs are computed simultaneously. Integration statements are placed at the end of the structure coding by the sorting algorithm, so that all current inputs to integration functions are defined before integration. A single routine is then used to update each of the integrator output variables used in the simulation.

Several different types of routines are available to perform the integration operation. They include both fixed step-size routines and variable integration step-size routines. Five fixed step-size routines are available: fixed Runge-Kutta, Simpson's, trapezoidal, rectangular, and second-order Adams. Two variable step-size routines are available: fifth-order Milne predictor-corrector and fourth-order Runge-Kutta. In the latter routines, the integration interval is automatically varied during problem execution to satisfy the user-specified error criterion.

If none of the above methods satisfies the user's requirement, a dummy integration routine named CENTRL can be used to specify a different integration method. The desired routine is entered into S/360 CSMP by giving it the name CENTRL.

The METHOD label specifies the particular integration routine to be used for the simulation. If none is specified, the RKS method is used. Names must be as shown below:

Name	Method
ADAMS	Second-order Adams integration with fixed interval
CENTRL	A dummy routine that may be replaced by a user-supplied centralized integration subroutine, if desired
MILNE	Variable-step, fifth-order, predictor-corrector Milne integration method
RECT	Rectangular integration
RKS	Fourth-order Runge-Kutta with variable integration interval; Simpson's Rule used for error estimation
RKSFX	Fourth-order Runge-Kutta with fixed interval
SIMP	Simpson's rule integration with fixed integration interval
TRAPZ	Trapezoidal integration

Many simulation models are most conveniently represented using subscripted variables with arrays of integrators. As an example, consider the phenomenon of unidirectional diffusion through some conducting medium. The dynamic behavior of such a system is described by a partial differential equation with both time and distance in the direction of diffusion as independent variables. To model such a system using S/360 CSMP, it must be approximated by a set of ordinary differential equations. To represent unidirectional diffusion, we view the conducting medium as though it were composed of a number of thin slices; with respect to the dependent variables of interest (perhaps temperature or concentration), each slice is considered homogeneous but gradients exist between adjacent slices to achieve the observed flow.

Since the computations are often identical for each of the slices, the use of subscripted variables in the equations is particularly appropriate. The "specification" form of the INTGRL statement permits the use of arrays for the inputs, outputs, and initial conditions of a group of integrators. As illustration, the statement

$$Z1 = INTGRL(ZIC1, DZDT1, 50)$$

specifies an array of 50 integrators and declares the symbolic names DZDT1, Z1, and ZIC1 to be the input, output, and initial condition for the first integrator of the array. Note especially that the third argument must be a literal integer constant. In order to gain access to the other elements of the arrays, the model should begin with the following DIMENSION and EQUIVALENCE statements:

/ DIMENSION DZDT(50), Z(50), ZIC(50)
/ EQUIVALENCE (DZDT1, DZDT(1)), (Z1, Z(1)), (ZIC1, ZIC(1))

While the normal restrictions of S/360 CSMP regarding the use of subscripts must still be observed, it is possible by such usage to compute the inputs and initial conditions through the subscripted variables DZDT and ZIC and to call upon the integrator outputs through the subscripted variable Z.

As an example, the statements shown below might be used to model diffusion through a conducting medium. The medium has been divided into 50 slices. The concentration Z in any compartment is obtained simply as the integral of the net flow through the compartment. For simplicity, both the upstream (SIDE 1) and the downstream (SIDE 2) concentrations are assumed constant. Flow into the nth compartment is assumed proportional to the difference $(Z_{n-1} - Z_n)$; flow out of the compartment, to $(Z_n - Z_{n+1})$. Note especially that subscripted quantities must be computed within a procedural or unsorted portion of the code. In this example, the derivative array DZDT is computed within a PROCEDURE so that full FORTRAN subscripting is permissible. The variable I is specified as FIXED so that it may be used as the index for a DO loop within the PROCEDURE. A similar technique might have been used within the INITIAL segment had it been desired to compute an initial condition array ZIC; since this was not done, the translator assumes that all elements of the initial condition array are equal to zero. Note also that it is not permissible to use a sub-scripted variable with the PRINT or PRTPLOT statement; thus, to print the output of the 15th compartment, a new variable Z15 is set equal to Z(15).

```
TITLE      ILLUSTRATION OF INTEGRATOR ARRAY
/    DIMENSION        DZDT(50), Z(50), ZIC(50)
/    EQUIVALENCE      (DZDT1, DZDT(1)), (Z1, Z(1)), (ZIC1, ZIC(1))
FIXED      I
INITIAL
PARAMETER      SIDE1 = 100.0, SIDE2 = 0.0, C = 2.3
DYNAMIC
Z1 = INTGRL(ZIC1, DZDT1, 50)
PROCEDURE      DZDT1 = DIFFUS(C, Z1, SIDE1, SIDE2)
     DZDT(1) = C * (SIDE1 − 2.0 * Z(1) + Z(2))
     DZDT(50) = C * (Z(49) − 2.0 * Z(50) + SIDE2)
     DO    5    I = 2, 49
5    DZDT(I) = C * (Z(I − 1) − 2.0 * Z(I) + Z(I + 1))
ENDPRO
NOSORT
Z15 = Z(15)
TIMER      DELT = 10.0, PRDEL = 5.0, OUTDEL = 5.0, FINTIM = 100.0
PRINT      SIDE1, Z15, SIDE2, DELT
END
STOP
ENDJOB
```

G. The MACRO Function

In some simulation problems, the selection of functional blocks available from the S/360 CSMP library may not be sufficient to describe conveniently the user's problem. The user, therefore, has been given means for building his own special purpose functional blocks. These functions may range from a few statements to an extremely complex model of a complete plant in a process control problem. To define special purpose functional blocks, either S/360 CSMP statements or FORTRAN or a combination of both may be used.

The MACRO type of function-defining capability of S/360 CSMP is a particularly powerful feature of the language. It allows the user to build larger functional blocks from the basic functions available in the library. Once defined, the MACRO may be used any number of times within the simulation structure statements.

As an illustration of this feature consider a simulation study that involves several functions with differing parameter values, but all having the general form

$$G(x) = a \sin^2(x) + b \sin(x) + c$$

The user may define a MACRO to represent this general functional relationship, assigning it some unique name, e.g., GUN. The CSMP statements to define this new MACRO might be as follows:

$$\text{MACRO}\quad \text{OUT} = \text{GUN(A, B, C, X)}$$
$$\text{T} = \text{SIN(X)}$$
$$\text{OUT} = \text{A} * \text{T} ** 2. + \text{B} * \text{T} + \text{C}$$
$$\text{ENDMAC}$$

Such MACRO definition cards must be placed at the beginning of the S/360 CSMP deck before any structure statements. Several rules that must be observed in defining MACRO functions may be found in the *S/360 CSMP User's Manual*.

H. The Structure of the Model

The main purpose of CSMP is to provide a computer mechanism for solving the differential equations that represent the dynamics of the model. However, there generally are computations that must be performed *before* each run and sometimes computations performed *after* each run. To satisfy these requirements, the general S/360 CSMP formulation of a model is divided into three segments:

1. The INITIAL segment is intended exclusively for computation of initial condition values and those parameters that the user prefers to express in terms of more basic parameters.

2. The DYNAMIC segment is normally the most extensive in the model, including the complete description of the system dynamics, together with any other computations desired during the run.

3. The TERMINAL segment is used for those computations desired after completion of each run. This will often be a simple calculation based on the final value of one or more model variables.

The explicit division of the model into computations to be performed before, during, and after each run is provided by the control statements: INITIAL, DYNAMIC, TERMINAL, and END.

I. Advantages of S/360 CSMP

The System/360 Continuous System Modeling Program provides the following advantages:

1. The input language is nonprocedural and free-form.
2. A problem can be prepared directly from a system of ordinary differential equations.
3. With few exceptions, FORTRAN statements can be intermixed with the S/360 CSMP simulation statements.
4. The method of integration can be chosen from several standard options provided in the program.
5. The output is provided automatically in a fixed format for all output options.
6. Thirty-four standard functional blocks are provided; in addition, the user can add his own functions to the library.
7. The entire S/360 CSMP simulation may be controlled by a sequence of conventional FORTRAN statements.

We have presented a condensed version of the *S/360 CSMP User's Manual* (H20-0367-2). A serious user must go to this manual in order to truly appreciate and take advantage of the multitude of features ignored in this presentation.

J. Sample Problem*

We consider the following two first-order irreversible reactions with rate constants k_1 and k_2:

$$A \xrightarrow{k_1} B \xrightarrow{k_2} C \tag{2.1}$$

If (A$_0$) denotes the initial concentration of A at time $t = 0$ and (B) = (C) = 0, the differential equations of the system are

$$\frac{dx}{dt} = -k_1 x \qquad \frac{dy}{dt} = k_1 x - k_2 y \qquad \frac{dz}{dt} = k_2 y \tag{2.2}$$

with $x = (A)/(A_0)$, $y = (B)/(A_0)$, and $z = (C)/(A_0)$. Integration gives

$$x = \exp(-k_1 t) \tag{2.3}$$

$$y = \frac{k_1}{k_2 - k_1} [\exp(-k_1 t) - \exp(-k_2 t)] \tag{2.4}$$

$$z = 1 - \frac{k_2}{k_2 - k_1} \exp(-k_1 t) + \frac{k_1}{k_2 - k_1} \exp(-k_2 t) \tag{2.5}$$

The concentration of A decreases monotonically while that of B first increases, reaches a maximum, and then decreases. The maximum is reached at $t_{max} = [1/(k_2 - k_1)] \ln(k_2/k_1)$, and the corresponding value of y is

$$y_{max} = \left(\frac{k_1}{k_2} \right)^{k_2/(k_2 - k_1)} \tag{2.6}$$

At t_{max}, the concentration of C versus t presents an inflection point, i.e., $d^2 z/dt^2 = 0$. The solution to this problem via S/360 CSMP is now presented. Figure 1 shows a complete listing of the statements that might be prepared for this problem. Figure 2 gives the fixed format tabular output from numerically integrating the differential equations and compares the results with the exact analytical solutions. Figures 3, 4, and 5 give the print-plot outputs.

The timing figures for the solution of this problem using CSMP on a S/360 Model 50 computer are as follows:

$$\text{Compiling time} = 32.22 \text{ sec}$$

$$\text{Execution time} = 3.78 \text{ sec}$$

$$\overline{\text{Total time} = 36.00 \text{ sec}}$$

---

* This problem is discussed by Professor Michel Boudart in *Kinetics of Chemical Processes*, Prentice-Hall, Inc., New Jersey, 1968, pp. 63–67.

```
*****************************************************************************
*                                                                         *
*                        REACTION KINETICS                                *
*              TWO FIRST-ORDER CONSECUTIVE REACTIONS                       *
*                                                                         *
*****************************************************************************


*       DESCRIPTION
*               REACTANT A GOES IRREVERSIBLY TO B WITH REACTION RATE
*               CONSTANT K1.  REACTANT B GOES TO C WITH RATE CONSTANT
*               K2.  RELATIVE CONCENTRATIONS ARE DEFINED AS: X=A/AO
*               Y=B/AO, Z=C/AO.  LET K1=2*K2.  LET THE INITIAL CONDI-
*               TIONS BE X=1.0, Y=0.0, Z=0.0.

        INITIAL
                      K1 = 2.*K2


        DYNAMIC

*           NUMERIC SOLUTION

*               EQUATIONS FOR DERIVITIVES:  DXDT, DYDT, DZDT
                      DXDT = -K1*X
                      DYDT = K1*X-K2*Y
                      DZDT = K2*Y

*               EQUATIONS FOR VARIABLES: X, Y, Z
                      X = INTGRL(XO,DXDT)
                      Y = INTGRL(YO,DYDT)
                      Z = INTGRL(ZO,DZDT)

*           THEORETICAL SOLUTIONS
                      XTH = EXP(-K1*TIME)
                      YTH = K1/(K2-K1)*(EXP(-K1*TIME)-EXP(-K2*TIME))
                      ZTH = 1-(K2*EXP(-K1*TIME)-K1*EXP(-K2*TIME))/(K2-K1)

*           INTEGRATION METHOD
                      METHOD RKS
*               (IF NO METHOD CARD IS USED, RKS METHOD IS ASSUMED.)

*           INITIAL AND FINISH CONDITIONS

*               INITIAL CONDITIONS
                      INCON     XO=1.
                      INCON     YO=0.
                      INCON     ZO=0.
                      PARAMETER K2=1.

*               FINISH CONDITIONS
                      TIMER FINTIM=5.0

*           OUTPUT

*               VARIABLES TO BE PRINTED AGAINST TIME
                      PRINT X,XTH,Y,YTH,Z,ZTH,DELT
                      TITLE   RELATIVE CONCENTRATIONS

*               PRINT INTERVAL
                      TIMER PRDEL=0.1

*               VARIABLES TO BE PLOTTED AGAINST TIME
                      PRTPLOT X
                      LABEL     REACTANT A ... RELATIVE CONCENTRATION
                      PRTPLOT Y
                      LABEL     REACTANT B ... RELATIVE CONCENTRATION
                      PRTPLOT Z
                      LABEL     REACTANT C ... RELATIVE CONCENTRATION

*               PLOT INTERVAL
                      TIMER OUTDEL=0.1

        END
        STOP
ENDJOB
```

Fig. 1. Computer listing of kinetics problem.

		RELATIVE CONCENTRATIONS				RKS	INTEGRATION
TIME	X	XTH	Y	YTH	Z	ZTH	DELT
0.0	1.0000E 00	1.0000E 00	0.0	0.0	0.0	0.0	6.2500E-03
1.0000E-01	8.1873E-01	8.1873E-01	1.7221E-01	1.7221E-01	9.0559E-03	9.0567E-03	5.6250E-02
2.0000E-01	6.7032E-01	6.7032E-01	2.9682E-01	2.9682E-01	3.2859E-02	3.2859E-02	1.0000E-01
3.0000E-01	5.4881E-01	5.4881E-01	3.8401E-01	3.8401E-01	6.7175E-02	6.7176E-02	1.0000E-01
4.0000E-01	4.4933E-01	4.4933E-01	4.4198E-01	4.4198E-01	1.0869E-01	1.0869E-01	1.0000E-01
5.0000E-01	3.6788E-01	3.6788E-01	4.7730E-01	4.7730E-01	1.5482E-01	1.5482E-01	1.0000E-01
6.0000E-01	3.0119E-01	3.0119E-01	4.9523E-01	4.9523E-01	2.0357E-01	2.0357E-01	1.0000E-01
7.0000E-01	2.4660E-01	2.4660E-01	4.9998E-01	4.9998E-01	2.5343E-01	2.5343E-01	1.0000E-01
8.0000E-01	2.0190E-01	2.0190E-01	4.9486E-01	4.9486E-01	3.0324E-01	3.0324E-01	1.0000E-01
9.0000E-01	1.6530E-01	1.6530E-01	4.8254E-01	4.8254E-01	3.5216E-01	3.5216E-01	1.0000E-01
1.0000E 00	1.3534E-01	1.3534E-01	4.6509E-01	4.6509E-01	3.9958E-01	3.9958E-01	1.0000E-01
1.1000E 00	1.1080E-01	1.1080E-01	4.4414E-01	4.4414E-01	4.4506E-01	4.4506E-01	1.0000E-01
1.2000E 00	9.0718E-02	9.0718E-02	4.2095E-01	4.2095E-01	4.8833E-01	4.8833E-01	9.9999E-02
1.3000E 00	7.4274E-02	7.4274E-02	3.9652E-01	3.9652E-01	5.2921E-01	5.2921E-01	1.0000E-01
1.4000E 00	6.0810E-02	6.0810E-02	3.7157E-01	3.7157E-01	5.6761E-01	5.6762E-01	9.9999E-02
1.5000E 00	4.9787E-02	4.9787E-02	3.4669E-01	3.4669E-01	6.0352E-01	6.0353E-01	1.0000E-01
1.6000E 00	4.0762E-02	4.0762E-02	3.2227E-01	3.2227E-01	6.3697E-01	6.3697E-01	1.0000E-01
1.7000E 00	3.3373E-02	3.3373E-02	2.9862E-01	2.9862E-01	6.6800E-01	6.6801E-01	9.9999E-02
1.8000E 00	2.7324E-02	2.7324E-02	2.7595E-01	2.7595E-01	6.9672E-01	6.9673E-01	1.0000E-01
1.9000E 00	2.2371E-02	2.2371E-02	2.5440E-01	2.5440E-01	7.2323E-01	7.2323E-01	9.9999E-02
2.0000E 00	1.8316E-02	1.8316E-02	2.3404E-01	2.3404E-01	7.4764E-01	7.4764E-01	1.0000E-01
2.1000E 00	1.4996E-02	1.4996E-02	2.1492E-01	2.1492E-01	7.7008E-01	7.7008E-01	9.9999E-02
2.2000E 00	1.2277E-02	1.2277E-02	1.9705E-01	1.9705E-01	7.9067E-01	7.9067E-01	1.0000E-01
2.3000E 00	1.0052E-02	1.0052E-02	1.8041E-01	1.8041E-01	8.0953E-01	8.0953E-01	9.9999E-02
2.4000E 00	8.2298E-03	8.2298E-03	1.6498E-01	1.6498E-01	8.2679E-01	8.2679E-01	1.0000E-01
2.5000E 00	6.7380E-03	6.7380E-03	1.5069E-01	1.5069E-01	8.4257E-01	8.4257E-01	1.0000E-01
2.6000E 00	5.5166E-03	5.5166E-03	1.3751E-01	1.3751E-01	8.5697E-01	8.5697E-01	9.9999E-02
2.7000E 00	4.5166E-03	4.5166E-03	1.2538E-01	1.2538E-01	8.7010E-01	8.7011E-01	1.0000E-01
2.8000E 00	3.6979E-03	3.6979E-03	1.1422E-01	1.1422E-01	8.8207E-01	8.8208E-01	9.9999E-02
2.9000E 00	3.0276E-03	3.0276E-03	1.0399E-01	1.0399E-01	8.9298E-01	8.9298E-01	1.0000E-01
3.0000E 00	2.4788E-03	2.4788E-03	9.4616E-02	9.4617E-02	9.0290E-01	9.0290E-01	9.9999E-02
3.1000E 00	2.0294E-03	2.0294E-03	8.6039E-02	8.6040E-02	9.1193E-01	9.1193E-01	1.0000E-01
3.2000E 00	1.6616E-03	1.6616E-03	7.8201E-02	7.8201E-02	9.2013E-01	9.2014E-01	1.0000E-01
3.3000E 00	1.3604E-03	1.3604E-03	7.1045E-02	7.1046E-02	9.2759E-01	9.2759E-01	9.9999E-02
3.4000E 00	1.1138E-03	1.1138E-03	6.4519E-02	6.4519E-02	9.3436E-01	9.3437E-01	1.0000E-01
3.5000E 00	9.1189E-04	9.1189E-04	5.8571E-02	5.8571E-02	9.4051E-01	9.4052E-01	9.9999E-02
3.6000E 00	7.4659E-04	7.4659E-04	5.3154E-02	5.3154E-02	9.4610E-01	9.4610E-01	1.0000E-01
3.7000E 00	6.1126E-04	6.1126E-04	4.8224E-02	4.8225E-02	9.5116E-01	9.5116E-01	9.9999E-02
3.8000E 00	5.0046E-04	5.0046E-04	4.3740E-02	4.3741E-02	9.5575E-01	9.5576E-01	1.0000E-01
3.9000E 00	4.0974E-04	4.0974E-04	3.9664E-02	3.9664E-02	9.5992E-01	9.5993E-01	9.9999E-02
4.0000E 00	3.3547E-04	3.3547E-04	3.5960E-02	3.5960E-02	9.6370E-01	9.6370E-01	1.0000E-01
4.1000E 00	2.7466E-04	2.7466E-04	3.2596E-02	3.2596E-02	9.6712E-01	9.6713E-01	1.0000E-01
4.2000E 00	2.2487E-04	2.2487E-04	2.9541E-02	2.9542E-02	9.7023E-01	9.7023E-01	9.9999E-02
4.3000E 00	1.8411E-04	1.8411E-04	2.6769E-02	2.6769E-02	9.7304E-01	9.7305E-01	1.0000E-01
4.4000E 00	1.5073E-04	1.5073E-04	2.4253E-02	2.4253E-02	9.7559E-01	9.7560E-01	9.9999E-02
4.5000E 00	1.2341E-04	1.2341E-04	2.1971E-02	2.1971E-02	9.7790E-01	9.7791E-01	1.0000E-01
4.6000E 00	1.0104E-04	1.0104E-04	1.9902E-02	1.9902E-02	9.7999E-01	9.8000E-01	9.9999E-02
4.7000E 00	8.2725E-05	8.2725E-05	1.8025E-02	1.8025E-02	9.8189E-01	9.8189E-01	1.0000E-01
4.8000E 00	6.7729E-05	6.7729E-05	1.6324E-02	1.6324E-02	9.8360E-01	9.8361E-01	1.0000E-01
4.9000E 00	5.5452E-05	5.5452E-05	1.4782E-02	1.4782E-02	9.8516E-01	9.8516E-01	9.9999E-02
5.0000E 00	4.5400E-05	4.5400E-05	1.3385E-02	1.3385E-02	9.8656E-01	9.8657E-01	1.0000E-01

Fig. 2. The S/360-CSMP fixed format tabular output (X, Y, Z) from the numerical integration of the kinetics differential equations and a comparison with the exact analytical solutions (XTH, YTH, ZTH).

REACTANT A ... RELATIVE CONCENTRATION

TIME	X	MINIMUM 4.5400E-05	X VERSUS TIME	MAXIMUM 1.0000E 00

```
                                MINIMUM                 X    VERSUS TIME         MAXIMUM
                                4.5400E-05                                       1.0000E 00
 TIME           X               I                                               I
 0.0            1.0000E 00      -----------------------------------------------------+
 1.0000E-01     8.1873E-01      ------------------------------------------------+
 2.0000E-01     6.7032E-01      ----------------------------------------+
 3.0000E-01     5.4881E-01      --------------------------------+
 4.0000E-01     4.4933E-01      ----------------------+
 5.0000E-01     3.6788E-01      ------------------+
 6.0000E-01     3.0119E-01      ---------------+
 7.0000E-01     2.4660E-01      ------------+
 8.0000E-01     2.0190E-01      ----------+
 9.0000E-01     1.6530E-01      --------+
 1.0000E 00     1.3534E-01      ------+
 1.1000E 00     1.1080E-01      -----+
 1.2000E 00     9.0718E-02      ----+
 1.3000E 00     7.4274E-02      ---+
 1.4000E 00     6.0810E-02      ---+
 1.5000E 00     4.9787E-02      --+
 1.6000E 00     4.0762E-02      --+
 1.7000E 00     3.3373E-02      -+
 1.8000E 00     2.7324E-02      -+
 1.9000E 00     2.2371E-02      -+
 2.0000E 00     1.8316E-02      +
 2.1000E 00     1.4996E-02      +
 2.2000E 00     1.2277E-02      +
 2.3000E 00     1.0052E-02      +
 2.4000E 00     8.2298E-03      +
 2.5000E 00     6.7380E-03      +
 2.6000E 00     5.5166E-03      +
 2.7000E 00     4.5166E-03      +
 2.8000E 00     3.6979E-03      +
 2.9000E 00     3.0276E-03      +
 3.0000E 00     2.4788E-03      +
 3.1000E 00     2.0294E-03      +
 3.2000E 00     1.6616E-03      +
 3.3000E 00     1.3604E-03      +
 3.4000E 00     1.1138E-03      +
 3.5000E 00     9.1189E-04      +
 3.6000E 00     7.4659E-04      +
 3.7000E 00     6.1126E-04      +
 3.8000E 00     5.0046E-04      +
 3.9000E 00     4.0974E-04      +
 4.0000E 00     3.3547E-04      +
 4.1000E 00     2.7466E-04      +
 4.2000E 00     2.2487E-04      +
 4.3000E 00     1.8411E-04      +
 4.4000E 00     1.5073E-04      +
 4.5000E 00     1.2341E-04      +
 4.6000E 00     1.0104E-04      +
 4.7000E 00     8.2725E-05      +
 4.8000E 00     6.7729E-05      +
 4.9000E 00     5.5452E-05      +
 5.0000E 00     4.5400E-05      +
```

Fig. 3. The relative concentration of reactant A versus time using the PRTPLT routine in S/360-CSMP.

```
                              MINIMUM              Y      VERSUS  TIME             MAXIMUM
                                0.0                                              4.9998E-01
       TIME          Y          I                                                   I
      0.0           0.0         +
      1.0000E-01    1.7221E-01  ----------------+
      2.0000E-01    2.9682E-01  ----------------------------+
      3.0000E-01    3.8401E-01  -------------------------------------+
      4.0000E-01    4.4198E-01  -------------------------------------------+
      5.0000E-01    4.7730E-01  ----------------------------------------------+
      6.0000E-01    4.9523E-01  ------------------------------------------------+
      7.0000E-01    4.9998E-01  -------------------------------------------------+
      8.0000E-01    4.9486E-01  ------------------------------------------------+
      9.0000E-01    4.8254E-01  -----------------------------------------------+
      1.0000E 00    4.6509E-01  ---------------------------------------------+
      1.1000E 00    4.4414E-01  -------------------------------------------+
      1.2000E 00    4.2095E-01  -----------------------------------------+
      1.3000E 00    3.9652E-01  --------------------------------------+
      1.4000E 00    3.7157E-01  ------------------------------------+
      1.5000E 00    3.4669E-01  ----------------------------------+
      1.6000E 00    3.2227E-01  -------------------------------+
      1.7000E 00    2.9862E-01  -----------------------------+
      1.8000E 00    2.7595E-01  --------------------------+
      1.9000E 00    2.5440E-01  ------------------------+
      2.0000E 00    2.3404E-01  -----------------------+
      2.1000E 00    2.1492E-01  ---------------------+
      2.2000E 00    1.9705E-01  -------------------+
      2.3000E 00    1.8041E-01  -----------------+
      2.4000E 00    1.6498E-01  ----------------+
      2.5000E 00    1.5069E-01  ---------------+
      2.6000E 00    1.3751E-01  -------------+
      2.7000E 00    1.2538E-01  ------------+
      2.8000E 00    1.1422E-01  -----------+
      2.9000E 00    1.0399E-01  ----------+
      3.0000E 00    9.4616E-02  ---------+
      3.1000E 00    8.6039E-02  --------+
      3.2000E 00    7.8201E-02  -------+
      3.3000E 00    7.1045E-02  -------+
      3.4000E 00    6.4519E-02  ------+
      3.5000E 00    5.8571E-02  -----+
      3.6000E 00    5.3154E-02  -----+
      3.7000E 00    4.8224E-02  ----+
      3.8000E 00    4.3740E-02  ----+
      3.9000E 00    3.9664E-02  ---+
      4.0000E 00    3.5960E-02  ---+
      4.1000E 00    3.2596E-02  ---+
      4.2000E 00    2.9541E-02  --+
      4.3000E 00    2.6769E-02  --+
      4.4000E 00    2.4253E-02  --+
      4.5000E 00    2.1971E-02  --+
      4.6000E 00    1.9902E-02  -+
      4.7000E 00    1.8025E-02  -+
      4.8000E 00    1.6324E-02  -+
      4.9000E 00    1.4782E-02  -+
      5.0000E 00    1.3385E-02  -+
```

Fig. 4. The relative concentration of reactant B versus time using the PRTPLT routine in S/360-CSMP.

```
                        REACTANT C ... RELATIVE CONCENTRATION

                        MINIMUM              Z     VERSUS TIME              MAXIMUM
                        0.0                                                 9.8656E-01
    TIME           Z        I                                                  I
   0.0           0.0        +
   1.0000E-01    9.0559E-03 +
   2.0000E-01    3.2859E-02 -+
   3.0000E-01    6.7175E-02 ---+
   4.0000E-01    1.0869E-01 -----+
   5.0000E-01    1.5482E-01 -------+
   6.0000E-01    2.0357E-01 ----------+
   7.0000E-01    2.5343E-01 ------------+
   8.0000E-01    3.0324E-01 ---------------+
   9.0000E-01    3.5216E-01 -----------------+
   1.0000E 00    3.9958E-01 --------------------+
   1.1000E 00    4.4506E-01 ----------------------+
   1.2000E 00    4.8833E-01 ------------------------+
   1.3000E 00    5.2921E-01 --------------------------+
   1.4000E 00    5.6761E-01 ----------------------------+
   1.5000E 00    6.0352E-01 -----------------------------+
   1.6000E 00    6.3697E-01 -------------------------------+
   1.7000E 00    6.6800E-01 --------------------------------+
   1.8000E 00    6.9672E-01 ----------------------------------+
   1.9000E 00    7.2323E-01 -----------------------------------+
   2.0000E 00    7.4764E-01 ------------------------------------+
   2.1000E 00    7.7008E-01 -------------------------------------+
   2.2000E 00    7.9067E-01 --------------------------------------+
   2.3000E 00    8.0953E-01 ---------------------------------------+
   2.4000E 00    8.2679E-01 ----------------------------------------+
   2.5000E 00    8.4257E-01 ----------------------------------------+
   2.6000E 00    8.5697E-01 -----------------------------------------+
   2.7000E 00    8.7010E-01 ------------------------------------------+
   2.8000E 00    8.8207E-01 ------------------------------------------+
   2.9000E 00    8.9298E-01 -------------------------------------------+
   3.0000E 00    9.0290E-01 -------------------------------------------+
   3.1000E 00    9.1193E-01 --------------------------------------------+
   3.2000E 00    9.2013E-01 --------------------------------------------+
   3.3000E 00    9.2759E-01 ---------------------------------------------+
   3.4000E 00    9.3436E-01 ---------------------------------------------+
   3.5000E 00    9.4051E-01 ---------------------------------------------+
   3.6000E 00    9.4610E-01 ----------------------------------------------+
   3.7000E 00    9.5116E-01 ----------------------------------------------+
   3.8000E 00    9.5575E-01 ----------------------------------------------+
   3.9000E 00    9.5992E-01 -----------------------------------------------+
   4.0000E 00    9.6370E-01 -----------------------------------------------+
   4.1000E 00    9.6712E-01 -----------------------------------------------+
   4.2000E 00    9.7023E-01 -----------------------------------------------+
   4.3000E 00    9.7304E-01 -----------------------------------------------+
   4.4000E 00    9.7559E-01 ------------------------------------------------+
   4.5000E 00    9.7790E-01 ------------------------------------------------+
   4.6000E 00    9.7999E-01 ------------------------------------------------+
   4.7000E 00    9.8189E-01 ------------------------------------------------+
   4.8000E 00    9.8360E-01 ------------------------------------------------+
   4.9000E 00    9.8516E-01 ------------------------------------------------+
   5.0000E 00    9.8656E-01 ------------------------------------------------+
```

Fig. 5. The relative concentration of reactant C versus time using PRTPLT routine
in S/360-CSMP.

III. HEAT TRANSFER IN AN INSULATED BAR

We consider the one-dimensional heat flow in a medium with thermal conductivity k, specific heat c, and density ϱ. The system of interest is a bar insulated on all sides and at one end (Fig. 6).

The bar initially has a uniform temperature of T_0. A temperature T_A is applied and maintained at the uninsulated end. We wish to find the temperature distribution of the bar as a function of time.

A. Finite Differencing the Heat Equation

The partial differential equation that relates the temperature to time and distance is the parabolic (heat) diffusion equation:

$$\varrho c \, \frac{\partial T}{\partial t} = \frac{\partial}{\partial x} \left(k \, \frac{\partial T}{\partial x} \right) \tag{3.1}$$

If k is independent of position in the bar, we obtain

$$\frac{\partial T}{\partial t} = \alpha \, \frac{\partial^2 T}{\partial x^2} \tag{3.2}$$

where $\alpha \equiv k/\varrho c \equiv$ thermal diffusivity. The boundary conditions are

$$T(x, t) = T_A \qquad \text{for } x = 0, \ t \geq 0 \tag{3.3}$$

$$\left(\frac{\partial T}{\partial x} \right) = 0 \qquad \text{for } x = L, \ t \geq 0 \tag{3.4}$$

The initial condition is

$$T(x, t) = T_0 \qquad \text{for } 0 \leq x \leq L \ \text{ and } \ t = 0 \tag{3.5}$$

We may solve the heat equation (3.2) by reducing the partial differential equation to a set of ordinary differential equations, which can then be solved by numerical methods. We apply finite differencing to the independent

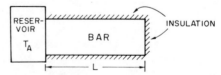

Fig. 6. An insulated bar in contact with a heat reservoir.

variable x of the partial differential equation and thus reduce the number of independent variables from two to one, i.e., $(x, t) \to (t)$.

When the finite difference is applied to a single space variable x, the maximum value of x, the length L of the bar, is divided into a discrete set of values of x:

$$x_1 = 0, \qquad x_2 = h, \qquad x_3 = 2h, \ldots, x_{N+1} = (N)h = L$$

Partial derivatives with respect to x are replaced by suitably chosen finite difference expressions at the respective values of x_i. When the partial derivative $\partial^2 T/\partial x^2$ is replaced by the central difference approximation,

$$\left(\frac{\partial^2 T}{\partial x^2} \right)_j \to \frac{1}{h} \left(\frac{T_{j+1} - T_j}{h} - \frac{T_j - T_{j-1}}{h} \right) \tag{3.6}$$

we obtain the set of ordinary differential equations

$$\frac{dT_j}{dt} = \frac{\alpha}{h^2} (T_{j+1} - 2T_j + T_{j-1}) \qquad j = 2, \ldots, N \tag{3.7}$$

where

$$T_j = T(x_j, t) \tag{3.8}$$

From the boundary condition (3.3), we note that

$$\frac{dT_1}{dt} = 0 \qquad \text{and} \qquad T_1 = T_A \tag{3.9}$$

Boundary condition (3.4) may be treated in the following way. From a Taylor's series expansion, we may write

$$T_N \cong T_{N+1} - h \left(\frac{\partial T}{\partial x} \right)_{N+1} + \frac{h^2}{2} \left(\frac{\partial^2 T}{\partial x^2} \right)_{N+1} \tag{3.10}$$

Since $(\partial T/\partial x)_{N+1} = 0$, we obtain, using Eq. (3.2),

$$T_N = T_{N+1} + \frac{h^2}{2} * \frac{1}{\alpha} \left(\frac{\partial T}{\partial t} \right)_{N+1}$$

or

$$\frac{dT_{N+1}}{dt} = \frac{2\alpha}{h^2} (T_N - T_{N+1}) \tag{3.11}$$

This equation is a *second-order* differential difference equation and would have been obtained by assuming a symmetry about the x_{N+1} plane; i.e., $T_{N+2} = T_{N-1}$.

The initial condition (3.5) for the differenced form of the problem is

$$T_j = T_0 \qquad j = 2, \ldots, N + 1 \quad \text{for } t = 0 \qquad (3.12)$$

Eqs. (3.7), (3.9), and (3.11) are the ordinary differential equations to be solved by numerical methods. We will now discuss the *model approximation* specified by these equations.

B. Finite Difference Approach in the Modeling

The finite difference approach in the modeling of this problem is to imagine that the bar is divided into a number of sections and that the temperature at the center of each section is the mean temperature of that section (Fig. 7).

Assume that each interior section is the same size h and the end sections are $h/2$. Then the rate of transfer from section $j - 1$ to section j is

$$J_{j-1} = \text{conductivity} * \text{area} * \text{temperature gradient}$$

$$J_{j-1} = k * A * \frac{T_{j-1} - T_j}{h} \qquad (3.13)$$

where A is the cross-sectional area of the bar, and linear temperature gradients are assumed between the centers of the sections.

Similarly, the rate of heat conduction from section j to section $j + 1$ is

$$J_j = k * A * \frac{T_j - T_{j+1}}{h} \qquad (3.14)$$

A heat balance made on section j results in the following differential equation:

Rate of heat accumulation = sum of all heat fluxes to the section

$$\frac{d}{dt}[\varrho c(Ah)T_j] = J_{j-1} - J_j \qquad (3.15)$$

or

$$\frac{d}{dt}T_j = \frac{\alpha}{h^2}[(T_{j-1} - T_j) + (T_{j+1} - T_j)] \qquad (3.16)$$

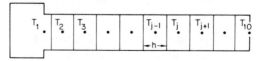

Fig. 7. The discretized model for the bar.

On the left-hand side of Eq. (3.15), we assumed that the "mean" cell temperature is T_j. We now examine the validity of this approximation.

The heat content of the jth cell is

$$\int_{j-\frac{1}{2}}^{j+\frac{1}{2}} \varrho c A T \, dx = \int_{j-\frac{1}{2}}^{j} \varrho c A \left[T_j - \frac{(T_{j-1} - T_j)}{h} (x - x_j) \right] dx$$

$$+ \int_{j}^{j+\frac{1}{2}} \varrho c A \left[T_j - \frac{(T_j - T_{j+1})}{h} (x - x_j) \right] dx$$

$$= \varrho c A h \left[T_j + \frac{h^2}{8} \frac{(T_{j+1} + T_{j-1} - 2T_j)}{h^2} \right]$$

In the limit as $h \rightarrow 0$,

$$\int \varrho c A T \, dx \rightarrow \varrho c A h \left[T + \frac{h^2}{8} \left(\frac{\partial^2 T}{\partial x^2} \right) \right] \qquad (3.17)$$

and the second term on the right-hand side of Eq. (3.17) is third order in h. Hence, the mean cell temperature approximation is an excellent one for small h.

Equations (3.16) were developed by the formal methods of finite differencing the heat diffusion (partial differential) equation, i.e., Eqs. (3.7). However, in many respects, the modeling approach developed here is both easier to visualize and straightforward to implement. Also, the actual physical model described by Eqs. (3.7) or (3.16) is clearly brought forth.

For the numerical solution of the differential equations via S/360-CSMP, we make the following choice for the parameters (units are cgs, calorie, °C):

$$T_A = 100°C$$
$$T_0 = 50°C$$
$$L = 10 \text{ cm}$$
$$k = 1.00$$
$$c = 0.0556 \left.\right\} \text{silver}$$
$$\varrho = 10.49$$

C. Fourier Solution

The Fourier series solution to this problem is

$$T(x, t) = T_A + \frac{2(T_0 - T_A)}{L} \sum_{n=1}^{\infty} M_n^{-1} e^{-\alpha M_n^2 t} \sin(M_n X) \qquad (3.18)$$

```
*****************************************************************************
*                                                                           *
*                       HEAT FLOW IN A METAL BAR                            *
*                                                                           *
*****************************************************************************

*       DESCRIPTION
*               A METAL BAR OF DENSITY RHO, SPECIFIC HEAT CP, AND THERMAL CONDUC-
*               TIVITY K IS HEATED AT ONE END.   THE REMAINING SURFACES ARE
*               INSULATED.   THE BAR IS DIVIDED INTO NS+1 GRID POINTS FOR NUMERICAL
*               INTEGRATION.   THE TEMPERATURE OF THE JTH GRID POINT IS TEMP(J).


*       DEFINITION OF VARIABLES
*               UNITS ARE CGS, CALORIE, DEGREE CENTIGRADE
*               TEMP(J)    = TEMPERATURE OF JTH GRID POINT
*               TEMPO(J)   = INITIAL TEMPERATURE OF JTH GRID POINT
*               RATE(J)    = RATE OF CHANGE OF TEMPERATURE AT JTH GRID POINT

*               NS         = NUMBER OF GRID POINTS USED FOR INTEGRATION
*               TO         = INITIAL TEMPERATURE OF ENTIRE BAR
*               TA         = TEMPERATURE OF RESERVOIR IN CONTACT WITH ONE END
*               LENGTH     = LENGTH OF BAR
*               CP         = SPECIFIC HEAT OF BAR
*               K          = THERMAL CONDUCTIVITY OF BAR
*               RHO        = DENSITY OF BAR
*               ALPHA      = THERMAL DIFFUSIVITY
*               H          = GRID POINT INTERVAL
*               THEO       = ONE IF FOURIER SERIES THEORETICAL SOLUTION DESIRED
*               N          = MAXIMUM NUMBER OF TERMS USED IN FOURIER SERIES
*               XPOS(I)    = POINTS ON BAR CHOSEN FOR FOURIER SERIES EVALUATION
*               OUT(I)     = TEMPERATURE AT ITH OUTPUT POSITION USING FOURIER
*                            SERIES SOLUTION

     INITIAL

                    NOSORT
                    FIXED I,J,N,NS,NSA,NF
                    NS = 10
                    DIMENSION TEMP(11), RATE(11), TEMPO(11)
                    DIMENSION XPOS(3), OUT(3)
*               PARAMETERS AND INITIAL CONDITIONS
*                 PARAMETERS FOR SILVER BAR
                    PARAMETER CP=0.0556
                    PARAMETER K=1.0
                    PARAMETER RHO=10.49
                    PARAMETER LENGTH=10.
                    PARAMETER TA=100.
                    PARAMETER THEO=1.
                    PARAMETER N=2000
                    INCON     TO=50.
                    CONSTANT  PI=3.14159

*               EVALUATION OF COEFFICIENTS

                    NSA      = NS+1
                    H        = LENGTH/NS
                    ALPHA    = K/(RHO*CP)
                    C        = ALPHA/(H*H)
                    XPOS(1)  = (NS/4)*LENGTH/NS
                    XPOS(2)  = (NS/2)*LENGTH/NS
                    XPOS(3)  = LENGTH
                    WRITE (6,21)  (XPOS(I),I=1,3)
         21         FORMAT(' POSITIONS OF OUTPUT POINTS ',5F10.5)

*               INITIAL TEMPERATURE
                    TEMPO(1) = TA
                    DO 30 J=2,NSA
                    TEMPO(J) = TO
         30         CONTINUE


     DYNAMIC

*          NUMERIC SOLUTION

*          CALCULATION OF TEMPERATURE CHANGE WITH TIME
                    NOSORT
                    DO 35 J=2,NS
                    RATE(J) = C*(TEMP(J-1)-2.*TEMP(J)+TEMP(J+1))
         35         CONTINUE
```

Fig. 8. Computer listing of heat diffusion problem.

```
*                          BOUNDARY CONDITIONS AS EXPRESSED BY RATE(1) AND RATE(NSA)
*                          SOURCE TEMPERATURE REMAINS CONSTANT
                           RATE(1)   = 0.

*                          FAR END OF BAR IS INSULATED
                           RATE(NSA) = C*(2.*TEMP(NS)-2.*TEMP(NSA))

*                       EQUATION FOR TEMPERATURE
*                          ARRAY INTGRL STATEMENTS REQUIRE THAT THE FIRST ARRAY POSITIONS
*                          BE RENAMED AS UNSUBSCRIPTED VARIABLES FOR INTEGRATION.
                           TEMP1 = INTGRL(TEMPO1,RATE1,11)
/                          EQUIVALENCE (TEMP1,TEMP(1)),(TEMPO1,TEMPO(1))
/                          EQUIVALENCE (RATE1,RATE(1))

*                       EQUATE ARRAY ELEMENTS TO UNSUBSCRIPTED VARIABLES FOR OUTPUT
                           IF (KEEP .NE. 1) GO TO 40
                           TOUT1 = TEMP(NS/4+1)
                           TOUT2 = TEMP(NS/2+1)
                           TOUT3 = TEMP(NSA)

*                       THEORETICAL SOLUTION
                           IF (THEO .NE. 1.) GO TO 40

                           DO 55 J=1,3
                           OUT(J) = 0.
      55                   CONTINUE
                           NFO = 0
*                       SUMMATION OF FOURIER SERIES
                           DO 45 I=1,N
                           M = (I-0.5)*PI/LENGTH
                           K1 = -ALPHA*TIME*M**2
                           IF (K1 .LE. -177.) GO TO 51
                           K1 = EXP(K1)
                           K2 = (TO-TA)*2./(M*LENGTH)
                           NF = 0
                           DO 50 J=1,3
                           TERM = K1*K2*SIN(M*XPOS(J))
*                          SERIES ENDS WHEN TERMS ARE LESS THAN .0005
                           IF(ABS(TERM) .LE. 5.E-4) NF=NF+1
                           OUT(J) = OUT(J)+TERM
                           IF (NF.NE.3) GO TO 50
                           NFO = I
                           GO TO 51
      50                   CONTINUE
      45                   CONTINUE

      51                   OUT1 = OUT(1)+TA
                           OUT2 = OUT(2)+TA
                           OUT3 = OUT(3)+TA
      40                   CONTINUE

*                    FINISH CONDITIONS
                        TIMER FINTIM=30.

*                    INTEGRATION METHOD
                        METHOD RKS


*              OUTPUT
*                    VARIABLES TO BE PRINTED AGAINST TIME
                        PRINT   TOUT1,OUT1,TOUT2,OUT2,TOUT3,OUT3,NFO
                        TITLE   TEMPERATURE OF SELECTED POSITIONS ON BAR

*                      PRINT INTERVAL
                        TIMER PRDEL=1.

*                    VARIABLES TO BE PLOTTED AGAINST TIME
                        PRTPLOT TOUT1,TOUT2,TOUT3
                        LABEL   TEMPERATURE OF SELECTED POSITIONS ON BAR

*                      PLOT INTERVAL
                        TIMER OUTDEL=1.


          END
          STOP
ENDJOB
```

Fig. 8 (*continued*).

where

$$M_n = \left(n - \frac{1}{2}\right) \frac{\pi}{L}$$

A comparison of the numerical solution with (3.18) will be made.

D. The S/360-CSMP Solution

The solution to this problem via S/360-CSMP in now presented. Figure 8 shows a complete listing of the statements that might be prepared for this problem. Figure 9 gives the fixed format tabular output from numerically integrating the differential equations and compares the results with the Fourier solution [Eq. (3.18)]. The print-plots are not shown.

The timing figures for the solution of this problem using CSMP on a S/360 Model 50 are as follows:

Compiling time = 42.47 sec

Execution time =   7.93 sec

Total = 50.40 sec

		TEMPERATURE OF SELECTED POSITIONS ON BAR				RKS	INTEGRATION
TIME	TOUT1	OUT1	TOUT2	OUT2	TOUT3	OUT3	NFO
0.0	5.0000E 01	5.0040E 01	5.0000E 01	5.0026E 01	5.0000E 01	5.0022E 01	0.0
1.0000E 00	6.4098E 01	6.4006E 01	5.0544E 01	5.0347E 01	5.0000E 01	5.0000E 01	8.0000E 00
2.0000E 00	7.2198E 01	7.2252E 01	5.3008E 01	5.2811E 01	5.0032E 01	5.0013E 01	6.0000E 00
3.0000E 00	7.6594E 01	7.6646E 01	5.6062E 01	5.5951E 01	5.0249E 01	5.0182E 01	5.0000E 00
4.0000E 00	7.9419E 01	7.9459E 01	5.8912E 01	5.8853E 01	5.0798E 01	5.0693E 01	5.0000E 00
5.0000E 00	8.1424E 01	8.1455E 01	6.1409E 01	6.1376E 01	5.1693E 01	5.1573E 01	4.0000E 00
6.0000E 00	8.2942E 01	8.2967E 01	6.3586E 01	6.3563E 01	5.2868E 01	5.2748E 01	4.0000E 00
7.0000E 00	8.4149E 01	8.4167E 01	6.5502E 01	6.5484E 01	5.4235E 01	5.4124E 01	4.0000E 00
8.0000E 00	8.5144E 01	8.5157E 01	6.7215E 01	6.7197E 01	5.5722E 01	5.5623E 01	4.0000E 00
9.0000E 00	8.5990E 01	8.5998E 01	6.8768E 01	6.8751E 01	5.7272E 01	5.7185E 01	4.0000E 00
1.0000E 01	8.6727E 01	8.6732E 01	7.0197E 01	7.0180E 01	5.8843E 01	5.8769E 01	3.0000E 00
1.1000E 01	8.7384E 01	8.7387E 01	7.1524E 01	7.1506E 01	6.0411E 01	6.0348E 01	3.0000E 00
1.2000E 01	8.7981E 01	8.7981E 01	7.2766E 01	7.2749E 01	6.1955E 01	6.1902E 01	3.0000E 00
1.3000E 01	8.8529E 01	8.8528E 01	7.3936E 01	7.3921E 01	6.3466E 01	6.3419E 01	3.0000E 00
1.4000E 01	8.9039E 01	8.9036E 01	7.5045E 01	7.5030E 01	6.4931E 01	6.4893E 01	3.0000E 00
1.5000E 01	8.9516E 01	8.9513E 01	7.6098E 01	7.6085E 01	6.6351E 01	6.6319E 01	3.0000E 00
1.6000E 01	8.9966E 01	8.9963E 01	7.7101E 01	7.7089E 01	6.7721E 01	6.7695E 01	3.0000E 00
1.7000E 01	9.0393E 01	9.0390E 01	7.8056E 01	7.8047E 01	6.9043E 01	6.9020E 01	3.0000E 00
1.8000E 01	9.0798E 01	9.0795E 01	7.8971E 01	7.8963E 01	7.0311E 01	7.0294E 01	3.0000E 00
1.9000E 01	9.1184E 01	9.1182E 01	7.9845E 01	7.9839E 01	7.1532E 01	7.1518E 01	3.0000E 00
2.0000E 01	9.1552E 01	9.1550E 01	8.0682E 01	8.0677E 01	7.2704E 01	7.2694E 01	3.0000E 00
2.1000E 01	9.1904E 01	9.1903E 01	8.1482E 01	8.1479E 01	7.3830E 01	7.3822E 01	3.0000E 00
2.2000E 01	9.2240E 01	9.2240E 01	8.2250E 01	8.2248E 01	7.4909E 01	7.4904E 01	3.0000E 00
2.3000E 01	9.2562E 01	9.2562E 01	8.2985E 01	8.2984E 01	7.5944E 01	7.5943E 01	3.0000E 00
2.4000E 01	9.2870E 01	9.2871E 01	8.3688E 01	8.3690E 01	7.6938E 01	7.6938E 01	3.0000E 00
2.5000E 01	9.3166E 01	9.3167E 01	8.4362E 01	8.4366E 01	7.7892E 01	7.7893E 01	3.0000E 00
2.6000E 01	9.3449E 01	9.3450E 01	8.5010E 01	8.5014E 01	7.8804E 01	7.8808E 01	3.0000E 00
2.7000E 01	9.3720E 01	9.3722E 01	8.5630E 01	8.5635E 01	7.9680E 01	7.9686E 01	3.0000E 00
2.8000E 01	9.3980E 01	9.3982E 01	8.6224E 01	8.6230E 01	8.0520E 01	8.0527E 01	2.0000E 00
2.9000E 01	9.4229E 01	9.4231E 01	8.6792E 01	8.6800E 01	8.1326E 01	8.1333E 01	2.0000E 00
3.0000E 01	9.4467E 01	9.4470E 01	8.7340E 01	8.7347E 01	8.2096E 01	8.2107E 01	2.0000E 00

Fig. 9. The S/360-CSMP fixed format tabular output (TOUT) from the numerical integration of the diffusion equation and a comparison with the Fourier solution (OUT).

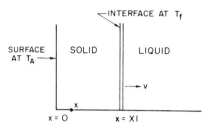

Fig. 10. The geometry of the freezing liquid.

IV. THE FREEZING OF A LIQUID*

A liquid is at a uniform initial temperature T_0. At a time $t = 0$, the temperature at the surface is lowered to T_A, and it is held constant for $t > 0$. The liquid freezes at a temperature T_f, where

$$T_A < T_f \leq T_0 \tag{4.1}$$

so that a frozen region grows into the liquid (see Figure 10).

We neglect convection currents and motion due to contraction or expansion upon freezing. Heat transfer is governed by the heat diffusion equations:

$$\alpha_S \frac{\partial^2 T_S}{\partial x^2} = \frac{\partial T_S}{\partial t} \quad \text{(solid)} \tag{4.2}$$

$$\alpha_L \frac{\partial^2 T_L}{\partial x^2} = \frac{\partial T_L}{\partial t} \quad \text{(liquid)} \tag{4.3}$$

where $\alpha = k/\varrho c$ is the thermal diffusivity, k is the thermal conductivity, ϱ is density, and c is specific heat.

Suppose L is the latent heat of fusion of the liquid and T_f is the melting point. Then if the surface of separation between the solid and liquid phases is at XI, one boundary condition to be satisfied at the interface is

$$T_S = T_L = T_f \quad \text{when } x = XI \tag{4.4}$$

A second condition concerns the liberation of latent heat at the interface. When the interface moves a distance $d(XI)$, a quantity of heat $\varrho L * d(XI)$ per unit area is liberated and must be removed by conduction.

---

* The description of the problem follows B. Carnahan, H. A. Luther, and J. O. Wilkes, *Applied Numerical Methods*, John Wiley and Sons, Inc., New York, 1969, pp. 524–525.

Table III. Some Thermal Parameters
for Ice and Water (cgs Units)

	Solid (ice)	Liquid (water)
k	0.0053	0.00144
α	0.0115	0.00144
$\varrho L = 73.6$		

The heat balance equation at the interface is

$$\varrho v L = k_S\left(\frac{\partial T_S}{\partial x}\right)_{x=XI^-} - k_L\left(\frac{\partial T_L}{\partial x}\right)_{x=XI^+} \tag{4.5}$$

which governs the velocity v of the interface.

Relevant data for the freezing of water are given in Table III.

A. Finite Differencing the Governing Equations

We now wish to derive the relevant differential-difference equations approximating the continuous representation of this problem for solution via S/360-CSMP. Figure 11 describes the grid structure for a finite difference description of the interface region.

Spatially differencing the heat diffusion equation for the solid and liquid regions results in the dependent variables $\{T_i\}$ and the independent variable t. In the heat balance equation (4.5) at the interface, we have the *additional dependent* variable XI where

$$v = \frac{d(XI)}{dt} \tag{4.6}$$

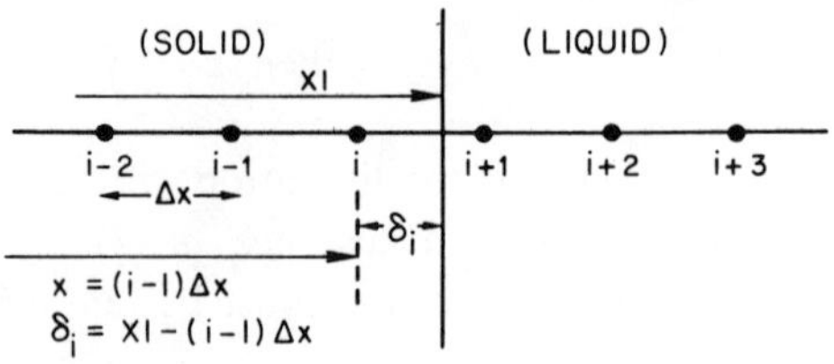

Fig. 11. The freezing interface moving between
two stationary grid points.

From Fig. 11 we write

$$T_i \cong T_f - \delta_i\left(\frac{\partial T}{\partial x}\right)_{XI^-} + \underbrace{\frac{\delta_i^2}{2}\left(\frac{\partial^2 T}{\partial x^2}\right)_{XI^-}}_{=0}$$

or

$$\left(\frac{\partial T}{\partial x}\right)_{XI^-} \cong \frac{T_f - T_i}{\delta_i} \tag{4.7}$$

Similarly,

$$\left(\frac{\partial T}{\partial x}\right)_{XI^+} \cong \frac{T_{i+1} - T_f}{\Delta x - \delta_i} \tag{4.8}$$

The interface heat balance Eq. (4.5) is approximated by

$$\frac{d(XI)}{dt} = \frac{1}{\varrho L}\left[k_S\left(\frac{T_f - T_i}{\delta_i}\right) - k_L\left(\frac{T_{i+1} - T_f}{\Delta x - \delta_i}\right)\right] \tag{4.9}$$

where special consideration is required to prevent numerical overflow when $\delta_i \cong 0$ or $\delta_i \cong \Delta x$.

The differencing of the heat diffusion equation requires special consideration in the region of the interface. Using Figure 11 and the Taylor expansion technique, we obtain the expressions·

$$T_{i-1} \cong T_i - \Delta x\left(\frac{\partial T}{\partial x}\right)_i + \frac{(\Delta x)^2}{2}\left(\frac{\partial^2 T}{\partial x^2}\right)_i$$

$$T_f \cong T_i + \delta_i\left(\frac{\partial T}{\partial x}\right)_i + \frac{\delta_i^2}{2}\left(\frac{\partial^2 T}{\partial x^2}\right)_i$$

From these two equations, we conclude that

$$\left(\frac{\Delta x}{2} + \frac{\delta_i}{2}\right)\left(\frac{\partial^2 T}{\partial x^2}\right)_i = \frac{T_{i-1} - T_i}{\Delta x} + \frac{T_f - T_i}{\delta_i} \tag{4.10}$$

From similar arguments, we obtain

$$\left(\frac{\Delta x - \delta_i}{2} + \frac{\Delta x}{2}\right)\left(\frac{\partial^2 T}{\partial x^2}\right)_{i+1} = \frac{T_f - T_{i+1}}{\Delta x - \delta_i} + \frac{T_{i+2} - T_{i+1}}{\Delta x} \tag{4.11}$$

The heat balance for grid points i and $i + 1$ (when $x_i \le XI \le x_{i+1}$) is

$$\frac{dT_i}{dt} = \frac{\alpha_S}{\Delta x/2 + \delta_i/2}\left(\frac{T_{i-1} - T_i}{\Delta x} - \frac{T_i - T_f}{\delta_i}\right) \tag{4.12}$$

$$\frac{dT_{i+1}}{dt} = \frac{\alpha_L}{(\Delta x - \delta_i)/2 + \Delta x/2}\left(\frac{T_f - T_{i+1}}{\Delta x - \delta_i} - \frac{T_{i+1} - T_{i+2}}{\Delta x}\right) \tag{4.13}$$

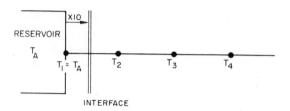

Fig. 12. The initial position of the interface at time
equal to zero.

where special consideration is required to prevent numerical overflow when
$\delta_i = 0$ or Δx. Otherwise the heat balance equations for the other grid
points are, for $j < i$,

$$\frac{dT_j}{dt} = \frac{\alpha_S}{(\Delta x)^2} (T_{j-1} + T_{j+1} - 2T_j) \qquad j = 2, \ldots, i-1 \quad (4.14)$$

for $j > i + 1$,

$$\frac{dT_j}{dt} = \frac{\alpha_L}{(\Delta x)^2} (T_{j-1} + T_{j+1} - 2T_j) \qquad j = i+2, \ldots \quad (4.15)$$

The differential equations to be numerically integrated are Eqs. (4.9), (4.14),
and (4.15).

The boundary conditions and initial conditions require special consideration also.

In order to get the interface explicitly positioned at $t = 0$, we place
it at point $XI0$ and the reservoir at grid point 1. Figure 12 depicts this
initial condition.

While we have pictured the liquid to be infinite in $+x$ extend, it is
impossible to realize this in a numerical calculation. If we assume $j = N$
to be the last grid point in our model, we note that our difference equations require a temperature at the nonexistent grid point $N + 1$. Any
number of approximations may be made to estimate T_{N+1}. The crudest
approximation is

$$T_{N+1} = T_N \qquad (4.16)$$

We will use this approximation. If we assumed that

$$T_{N+1} = T_{N-1} \qquad (4.17)$$

```
*****************************************************************************
*                                                                           *
*                       MOVING SOLID-LIQUID INTERFACE                       *
*                                                                           *
*****************************************************************************
*       DESCRIPTION
*           WATER AT AN INITIAL TEMPERATURE TO IS COVERED WITH A LAYER OF ICE.
*           THE UPPER SURFACE OF THE ICE IS MAINTAINED AT A TEMPERATURE
*           TA.  THE ICE-WATER INTERFACE IS LOCATED AT A DEPTH XI WITH A
*           TEMPERATURE TF AND MOVES DOWNWARD AS THE WATER FREEZES.  THE
*           SAMPLE IS DIVIDED INTO NS+1 GRID POINTS FOR INTEGRATION.
*           THE TEMPERATURE OF THE JTH GRID POINT IS TEMP(J).
*           DEFINITION OF VARIABLES
*               UNITS CGS, CALORIE, DEGREE CENTIGRADE
*               TEMP(J)   = TEMPERATURE OF JTH GRID POINT
*               TEMPO(J)  = INITIAL TEMPERATURE OF JTH GRID POINT
*               RATE(J)   = RATE OF CHANGE OF TEMPERATURE AT JTH GRID POINT
*           THERMAL PROPERTIES OF SOLID  (ICE)
*               RHOS      = DENSITY
*               KS        = THERMAL CONDUCTIVITY
*               CPS       = SPECIFIC HEAT
*               ALPHAS    = THERMAL DIFFUSIVITY
*           THERMAL PROPERTIES OF LIQUID  (WATER)
*               RHOL      = DENSITY
*               KL        = THERMAL CONDUCTIVITY
*               CPL       = SPECIFIC HEAT
*               ALPHAL    = THERMAL DIFFUSIVITY
*               HEATF     = LATENT HEAT OF FUSION PER UNIT VOLUME
*               NS        = NUMBER OF GRID POINTS USED FOR INTEGRATION
*               TA        = TEMPERATURE OF RESERVOIR IN CONTACT WITH SOLID
*                           AT ZERO TIME
*               TO        = INITIAL TEMPERATURE OF LIQUID
*               TF        = FREEZING TEMPERATURE OF LIQUID
*               LENGTH    = LENGTH OF SAMPLE
*               H         = GRID POINT INTERVAL
*               SIGMA     = DISTANCE FROM LAST GRID POINT IN SOLID TO
*                           INTERFACE
*               XIO       = INITIAL POSITION OF INTERFACE
*               XI        = POSITION OF INTERFACE IN SAMPLE
*               VEL       = VELOCITY OF INTERFACE
*               I         = INDEX OF LAST GRID POINT IN SOLID
*               POSI      = POSITION OF LAST GRID POINT IN SOLID
*               GRADS     = TEMPERATURE GRADIENT IMMEDIATELY ON SOLID SIDE
*                           OF INTERFACE
*               GRADL     = TEMPERATURE GRADIENT IMMEDIATELY ON LIQUID SIDE
*                           OF INTERFACE
        INITIAL

                    NOSORT
                    NS = 10
                    DIMENSION  TEMP(12), RATE(12), TEMPO(12)
                    FIXED IZ,I,IA,IB,NS,NSA,J,NSB

*           PARAMETERS AND INITIAL CONDITIONS
*               PARAMETERS FOR SOLID  (ICE)
                    PARAMETER RHOS=0.92
                    PARAMETER KS=0.0053
                    PARAMETER CPS=0.50095

*               PARAMETERS FOR LIQUID (WATER)
                    PARAMETER RHOL=1.0
                    PARAMETER KL=0.00144
                    PARAMETER CPL=1.0

                    PARAMETER HEATF=73.6
                    PARAMETER LENGTH=1.0
                    PARAMETER TA=-3.0
                    PARAMETER TO=2.0
                    PARAMETER TF=0.

*               EVALUATION OF COEFFICIENTS
                    NSA      = NS+1
                    NSB      = NS+2
                    H        = LENGTH/NS
                    LAST     = (NS-1)*H
                    XIO      = H*.02
                    ALPHAS   = KS/(RHOS*CPS)
                    CS       = ALPHAS/(H*H)
                    ALPHAL   = KL/(RHOL*CPL)
                    CL       = ALPHAL/(H*H)

*               INITIAL TEMPERATURE
*                   INTERFACE INITIALLY LOCATED AT XIO = H*0.02
*                   ICE EXTENDS FROM XI = 0.  TO XI = XIO
```

Fig. 13. Computer listing of the freezing interface problem.

```
*              WATER IS OF INFINITE EXTENT IN THE X DIRECTION
*              (SIMULATION ENDS AT XI = LENGTH.)
                  TEMPO(1) = TA
                  DO 30 J=2,NSB
                  TEMPO(J) = TO
    30            CONTINUE
                  XI = XIO


        DYNAMIC

*          NUMERIC SOLUTION

*          CALCULATION OF TEMPERATURE GRADIENTS AT THE INTERFACE
                  NOSORT
                  I      = XI/H+1.0001
                  IZ     = I-1
                  IA     = I+1
                  IB     = I+2
                  POSI   = (I-1)*H
                  SIGMA  = XI-POSI

*              WHEN THE INTERFACE IS CLOSE TO GRID POINT I, THE GRADIENT
*              ON THE SOLID SIDE OF THE INTERFACE IS COMPUTED FROM TEMP(I-1)
*              INSTEAD OF TEMP(I).
                  IF(SIGMA.GE.H*.01) GO TO 31
                  GRADS = (TF-TEMP(I-1))/(H+SIGMA)
                  GRADL = (TEMP(I+1)-TF)/(H-SIGMA)
                  GO TO 33

*              WHEN THE INTERFACE IS CLOSE TO GRID POINT I+1, THE GRADIENT
*              ON THE LIQUID SIDE OF THE INTERFACE IS COMPUTED FROM TEMP(I+2)
*              INSTEAD OF TEMP(I+1).
    31            IF(SIGMA.LE.H*.99) GO TO 32
                  GRADS = (TF-TEMP(I))/SIGMA
                  GRADL = (TEMP(I+2)-TF)/(2.*H-SIGMA)
                  GO TO 33

*              INTERFACE IS BETWEEN GRID POINTS I AND I+1.  GRADIENTS
*              ARE CALCULATED NORMALLY.
    32            GRADS = (TF-TEMP(I))/SIGMA
                  GRADL = (TEMP(I+1)-TF)/(H-SIGMA)
    33            IF(I.EQ.2) GO TO 34
                  IF(I.EQ.1) GO TO 37


*          CALCULATION OF TEMPERATURE CHANGE WITH TIME

*              TEMPERATURE RATE IN SOLID FOR GRID POINTS NOT NEXT TO INTERFACE
                  DO 35 J=2,IZ
                  RATE(J) = CS*(TEMP(J-1)-2.*TEMP(J)+TEMP(J+1))
    35            CONTINUE

*              TEMPERATURE RATE FOR LAST GRID POINT IN SOLID
    34            RATE(I)   = ALPHAS/(0.5*(H+SIGMA))* ...
                            (GRADS-(TEMP(I)-TEMP(I-1))/H)

*              TEMPERATURE RATE FOR FIRST GRID POINT IN LIQUID
    37            RATE(I+1) = ALPHAL/(H-0.5*SIGMA)* ...
                            ((TEMP(I+2)-TEMP(I+1))/H-GRADL)
                  IF(I.GE.NS) GO TO 43

*              TEMPERATURE RATE IN LIQUID FOR GRID POINTS NOT NEXT TO INTERFACE
                  DO 36 J=IB,NSA
                  RATE(J) = CL*(TEMP(J-1)-2.*TEMP(J)+TEMP(J+1))
    36            CONTINUE


*              BOUNDARY CONDITIONS AS EXPRESSED BY RATE(1) AND RATE(NSA)
*              SOURCE TEMPERATURE IS CONSTANT
    43            RATE(1) = 0.0

*              TO SIMULATE AN INFINITE EXTENT OF LIQUID, WE ASSUME
*              TEMP(N+1) = TEMP(N).
                  RATE(NSA+1) = RATE(NSA)


*          CALCULATION OF INTERFACE VELOCITY
                  VEL = (KS*GRADS-KL*GRADL)/HEATF

*          EQUATION FOR TEMPERATURE
```

Fig. 13 *(continued)*.

```
*                     ARRAY INTGRL STATEMENTS REQUIRE THAT THE FIRST ARRAY POSITIONS
*                     BE RENAMED AS UNSUBSCRIPTED VARIABLES FOR INTEGRATION.
/                     TEMP1 = INTGRL(TEMPO1,RATE1,12)
/                     EQUIVALENCE (TEMP1,TEMP(1)),(TEMPO1,TEMPO(1))
                      EQUIVALENCE (RATE1,RATE(1))

*           EQUATION FOR POSITION OF INTERFACE
                      XI = INTGRL(XIO,VEL)

*              INTEGRATION METHOD
                      METHOD RKS


*        OUTPUT

*           EQUATE ARRAY ELEMENTS TO UNSUBSCRIPTED VARIABLES FOR OUTPUT
                      TEMP2  = TEMP(2)
                      TEMP3  = TEMP(3)
                      TEMP4  = TEMP(4)
                      TEMP5  = TEMP(5)
                      TEMP6  = TEMP(6)
                      TEMP7  = TEMP(7)
                      TEMP8  = TEMP(8)
                      TEMP9  = TEMP(9)
                      TEMP10 = TEMP(10)
        40            CONTINUE


*           VARIABLES TO BE PRINTED AGAINST TIME
                      PRINT TEMP2,TEMP4,TEMP6,TEMP10,XI,VEL
*              PRINT INTERVAL
                      TIMER PRDEL=50.0


*           VARIABLES TO BE PLOTTED AGAINST TIME
                      PRTPLOT TEMP2,TEMP3,TEMP4,TEMP5,TEMP6
                      PRTPLOT TEMP7,TEMP8,TEMP9,TEMP10
                      LABEL TEMPERATURE IN THE MOVING INTERFACE PROBLEM
                      LABEL TEMPERATURE IN THE MOVING INTERFACE PROBLEM

*              PLOT INTERVAL
                      TIMER OUTDEL=50.0


*           FINISH CONDITIONS
*           SIMULATION HALTS WHEN INTERFACE REACHES LAST GRID POINT
                      FINISH XI=LAST
                      TIMER FINTIM=2500.0

            END
            STOP
ENDJOB
```

Fig. 13 (*continued*).

we would be describing a one-dimensional "ice cube" with the symmetry plane at grid point N.

B. The S/360-CSMP Solution

The solution to this problem via S/360-CSMP is now presented. Figure 13 shows a complete listing of the statements that might be prepared for this problem. Figure 14 gives the fixed-format tabular output from the numerical integration (TEMP, XI, VEL). Selected print-plots of the temperature versus time at various grid points are presented in Figs. 15 and 16.

						RKS	INTEGRATION
TIME	TEMP2	TEMP4	TEMP6	TEMP10	XI	VEL	
0.0	2.0000E 00	2.0000E 00	2.0000E 00	2.0000E 00	2.0000E-03	1.0762E-01	
5.0000E 01	-8.8131E-01	7.8638E-01	1.4654E 00	1.9399E 00	1.4185E-01	1.4121E-03	
1.0000E 02	-1.4983E 00	3.7216E-01	9.9784E-01	1.6586E 00	2.0039E-01	1.0016E-03	
1.5000E 02	-1.7740E 00	1.6362E-01	6.9868E-01	1.3223E 00	2.4532E-01	8.1803E-04	
2.0000E 02	-1.9386E 00	3.9506E-02	4.7977E-01	1.0145E 00	2.8339E-01	7.1303E-04	
2.5000E 02	-2.0516E 00	-1.6127E-01	3.1740E-01	7.5421E-01	3.1715E-01	6.4177E-04	
3.0000E 02	-2.1354E 00	-4.1117E-01	2.0028E-01	5.4376E-01	3.4792E-01	5.9124E-04	
3.5000E 02	-2.2009E 00	-6.0670E-01	1.1952E-01	3.7981E-01	3.7648E-01	5.5183E-04	
4.0000E 02	-2.2540E 00	-7.6539E-01	6.6489E-02	2.5653E-01	4.0331E-01	5.1855E-04	
4.5000E 02	-2.2980E 00	-8.9672E-01	3.3556E-02	1.6747E-01	4.2857E-01	4.9196E-04	
5.0000E 02	-2.3352E 00	-1.0080E 00	1.4658E-02	1.0539E-01	4.5258E-01	4.6876E-04	
5.5000E 02	-2.3672E 00	-1.1037E 00	4.7893E-03	6.3748E-02	4.7549E-01	4.4829E-04	
6.0000E 02	-2.3951E 00	-1.1872E 00	3.0453E-04	3.6971E-02	4.9745E-01	4.2999E-04	
6.5000E 02	-2.4198E 00	-1.2609E 00	-1.0683E-01	2.0429E-02	5.1859E-01	4.1251E-04	
7.0000E 02	-2.4416E 00	-1.3263E 00	-2.1519E-01	1.0765E-02	5.3887E-01	3.9791E-04	
7.5000E 02	-2.4612E 00	-1.3847E 00	-3.1220E-01	5.3813E-03	5.5842E-01	3.8442E-04	
8.0000E 02	-2.4788E 00	-1.4376E 00	-3.9980E-01	2.5403E-03	5.7733E-01	3.7211E-04	
8.5000E 02	-2.4948E 00	-1.4855E 00	-4.7942E-01	1.1273E-03	5.9565E-01	3.6085E-04	
9.0000E 02	-2.5095E 00	-1.5295E 00	-5.5239E-01	4.6195E-04	6.1348E-01	3.4977E-04	
9.5000E 02	-2.5229E 00	-1.5697E 00	-6.1910E-01	1.7734E-04	6.3074E-01	3.4049E-04	
1.0000E 03	-2.5353E 00	-1.6067E 00	-6.8064E-01	6.2674E-05	6.4754E-01	3.3172E-04	
1.0500E 03	-2.5468E 00	-1.6411E 00	-7.3766E-01	2.0227E-05	6.6393E-01	3.2363E-04	
1.1000E 03	-2.5574E 00	-1.6730E 00	-7.9068E-01	5.9169E-06	6.7992E-01	3.1609E-04	
1.1500E 03	-2.5674E 00	-1.7028E 00	-8.4017E-01	1.5533E-06	6.9554E-01	3.0904E-04	
1.2000E 03	-2.5767E 00	-1.7307E 00	-8.8653E-01	3.8238E-07	7.1085E-01	3.0193E-04	
1.2500E 03	-2.5854E 00	-1.7568E 00	-9.2995E-01	7.6961E-08	7.2581E-01	2.9587E-04	
1.3000E 03	-2.5936E 00	-1.7814E 00	-9.7077E-01	1.3233E-08	7.4045E-01	2.9001E-04	
1.3500E 03	-2.6013E 00	-1.8044E 00	-1.0092E 00	1.9086E-09	7.5481E-01	2.8456E-04	
1.4000E 03	-2.6086E 00	-1.8264E 00	-1.0457E 00	2.2632E-10	7.6892E-01	2.7943E-04	
1.4500E 03	-2.6156E 00	-1.8473E 00	-1.0804E 00	2.1623E-11	7.8276E-01	2.7462E-04	
1.5000E 03	-2.6221E 00	-1.8668E 00	-1.1129E 00	-1.5897E-09	7.9637E-01	2.6989E-04	
1.5500E 03	-2.6284E 00	-1.8856E 00	-1.1440E 00	-2.5070E-08	8.0977E-01	2.6512E-04	
1.6000E 03	-2.6343E 00	-1.9034E 00	-1.1736E 00	-9.6794E-10	8.2292E-01	2.6093E-04	
1.6500E 03	-2.6400E 00	-1.9203E 00	-1.2019E 00	-2.4181E-11	8.3587E-01	2.5694E-04	
1.7000E 03	-2.6454E 00	-1.9366E 00	-1.2288E 00	-3.6466E-13	8.4861E-01	2.5307E-04	
1.7500E 03	-2.6505E 00	-1.9520E 00	-1.2547E 00	-3.0264E-15	8.6118E-01	2.4923E-04	
1.8000E 03	-2.6555E 00	-1.9669E 00	-1.2793E 00	-1.2294E-17	8.7356E-01	2.4596E-04	
1.8500E 03	-2.6602E 00	-1.9810E 00	-1.3029E 00	-1.9936E-20	8.8577E-01	2.4262E-04	
1.9000E 03	-2.6648E 00	-1.9948E 00	-1.3257E 00	-9.1379E-05	8.9782E-01	2.4018E-04	

0***SIMULATION HALTED***   XI   = 9.0001E-01

1.9092E 03 -2.6656E 00 -1.9972E 00 -1.3298E 00 -1.2732E-05 9.0001E-01 2.3877E-04

Fig. 14. The S/360-CSMP fixed format tabular output (TEMP, XI, VEL) from the numerical integration.

```
                    TEMPERATURE IN THE MOVING INTERFACE PROBLEM

                            MINIMUM            TEMP7   VERSUS TIME          MAXIMUM
                           -9.9656E-01                                     2.0000E 00
       TIME         TEMP7        I                                             I
       0.0          2.0000E 00   --------------------------------------------------+
       5.0000E 01   1.6724E 00   ------------------------------------------------+
       1.0000E 02   1.2384E 00   ---------------------------------------------+
       1.5000E 02   9.1645E-01   ----------------------------------------+
       2.0000E 02   6.6367E-01   ---------------------------------+
       2.5000E 02   4.6616E-01   ----------------------------+
       3.0000E 02   3.1624E-01   ----------------------+
       3.5000E 02   2.0663E-01   --------------------+
       4.0000E 02   1.2953E-01   ------------------+
       4.5000E 02   7.7565E-02   ----------------+
       5.0000E 02   4.4169E-02   ----------------+
       5.5000E 02   2.3757E-02   ----------------+
       6.0000E 02   1.1959E-02   ---------------+
       6.5000E 02   5.5085E-03   ---------------+
       7.0000E 02   2.2847E-03   ---------------+
       7.5000E 02   8.1433E-04   ---------------+
       8.0000E 02   2.1983E-04   ---------------+
       8.5000E 02   2.0825E-05   ---------------+
       9.0000E 02  -6.5503E-02   ---------------+
       9.5000E 02  -1.4537E-01   --------------+
       1.0000E 03  -2.1901E-01   ------------+
       1.0500E 03  -2.8730E-01   -----------+
       1.1000E 03  -3.5079E-01   -----------+
       1.1500E 03  -4.1004E-01   ---------+
       1.2000E 03  -4.6556E-01   --------+
       1.2500E 03  -5.1755E-01   -------+
       1.3000E 03  -5.6646E-01   -------+
       1.3500E 03  -6.1261E-01   ------+
       1.4000E 03  -6.5619E-01   -----+
       1.4500E 03  -6.9735E-01   ----+
       1.5000E 03  -7.3667E-01   ----+
       1.5500E 03  -7.7402E-01   ---+
       1.6000E 03  -8.0947E-01   ---+
       1.6500E 03  -8.4329E-01   --+
       1.7000E 03  -8.7558E-01   --+
       1.7500E 03  -9.0634E-01   -+
       1.8000E 03  -9.3609E-01   -+
       1.8500E 03  -9.6453E-01   +
       1.9000E 03  -9.9169E-01   +
       1.9092E 03  -9.9656E-01   +
```

Fig. 15. TEMP7 versus time. Note the plateau where the temperature at grid point # 7 has reached zero degrees centigrade but the interface has not yet passed.

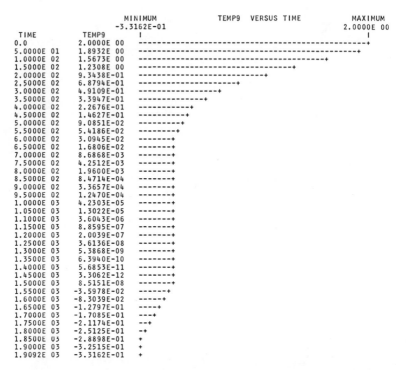

Fig. 16. TEMP9 versus time. Note the plateau where the temperature at grid point # 9
has reached zero degrees centigrade but the interface has not yet passed.

The timing figures for the solution of this problem using CSMP on a
S/360 Model 50 computer are as follows:

Compiling time =   54 sec

Execution time = 244 sec

Total = 298 sec

Chapter 4

Computer Simulation of Vapor Deposition on Two-Dimensional Lattices

George M. White

Xerox Palo Alto Research Center
Palo Alto, California

I. BASIC CONCEPTS OF PHYSICAL PROCESSES

A. Introduction

Abraham and White[1,16] have written computer programs that allow a user to run "computer experiments" for vapor deposition studies that include first and second nearest neighbor interactions. These Vapor Deposition Simulation programs (VDS for short) use Monte Carlo methods to determine the molecular dynamics of condensation, evaporation, and migration on lattices. Results are expressed in terms of adsorption isotherms and lattice coverages as a function of time.

Vapor deposition of molecules on lattice networks has been extensively studied.[2,4] In the past, theoretical treatments have failed to realistically incorporate the effects of nearest neighbor and next-nearest neighbor interactions between adsorbed molecules. However, Honig[3] has recently made significant advances by using a complicated counting scheme that considers occupation configurations of neighboring lattice sites in detail. The theoretical results of Honig are difficult to verify experimentally and so have remained without experimental corroboration until now.

The simulation methods employed in the VDS program are explained in detail in Sections V and VII. The principal application has been to systems that permit a comparison with Honig's work, although the simulation methods are easily applied to a variety of other problems. The results of the simulations demonstrate that Honig's treatment is quite accurate and is a substantial improvement over previous treatments. This agreement also serves to build confidence in the use of Monte Carlo methods in simulating molecular dynamics for vapor deposition studies.

Pioneering studies in adsorption on crystal surfaces using Monte Carlo methods include Moore,[12] Gilmer and Bennema,[13] and Gordon.[9]

B. The Honig Model

Honig's theory is based on a simplified model of a crystalline surface. Honig assumes the following:

1. Adsorption occurs on a fixed number of definite sites to which adsorbed atoms are bound during residence on the lattice.
2. Each site has the same binding energy for adsorbed atoms.
3. The coverage is limited to monolayer.
4. The spacing between lattice sites is greater than the greatest linear extension of the adsorbate atoms so that one site suffices for the accommodation of one atom on the surface.
5. The gas above the lattice consists of a single species.
6. Edge effects may be ignored.

Two basic lattice representations are of interest, square lattices and hexagonal lattices, but we will only discuss the former in detail. For square lattices the four next nearest neighbors are located on opposite corners of the square formed by the four nearest neighbors when each nearest neighbor is in the middle of one side of the square. The ratio of distances separating the center of the square from the next nearest and nearest neighbors is $r_2/r_1 = \sqrt{2}$. The interaction forces are assumed to fall off as the sixth power of the distance between adsorbate atoms as predicted by van der Walls–London dispersion forces. It follows that $w_2/w_1 = \frac{1}{8}$ where w is the interaction energy.

Honig's approach rests on the methodology of order–disorder theory and, in particular, a configuration counting scheme originally developed by Hijmans and deBoer.[5] Using a statistical mechanical approach, Honig addresses himself to the problem of calculating the adsorption isotherms, i.e., specifying the fraction of all surface sites occupied by the adsorbate at a given temperature and gas pressure, taking lateral interactions between first and second nearest neighbors into account. His results differ significantly from those of the Fowler–Guggenheim theory, which also takes lateral interactions into account. In our computer simulation approach to the problem, we obtain equilibrium adsorption isotherms identical with those of Honig. We also obtain the time-dependent development of the lattice coverage.

C. The Simulated Processes

The dynamical processes simulated are condensation, evaporation, and migration of vapor molecules on the lattice surface. The system is considered to be Markovian, i.e., the transition that the system undergoes at time t is dependent only upon the state of the system at $t - 1$ and not $t - 2$ or $t - 3$, etc. The processes are quantized into "all or nothing" transitions. For example, a molecule may migrate from an initial site to a nearest neighbor in any given time quantum. It may never stop midway between sites. Furthermore, only one transition may occur per time quantum. If an evaporation occurs, i.e., an adsorbed molecule breaks its bonds and reenters the gas phase, no migrations or condensations are permitted during the same time interval. This restriction to a maximum of one transition per time quantum is a standard mathematical assumption in simulating dynamic systems. However, allowing transitions that produce large changes in geometrical configurations in a single time quantum may not always be valid. Our model assumes ideal gas behavior: particles that are not absorbed to the surface do not interact in any way. If the effects of nonideal gases were to be studied, shorter spacial transitions would be required.

The following fundamental atomic processes are assumed to be the only important mechanisms for changing the state of the system at any given time:

1. *Condensation* of a vapor atom on a "crystalline" site or "deposited-layer" site (vapor atom → adatom).
2. *Evaporation* of an adatom to the vapor (adatom → vapor atom).
3. *Migration* of the adatom to a neighboring site (adatom → adatom).

The pertinent rate equations for these processes will now be explained.

D. The Rate Equations

The rate at which condensation occurs is directly proportional to the rate at which atoms strike the surface. The number α of vapor atoms striking a unit area of a surface per unit time is given by

$$\alpha = \frac{P}{(2\pi mkT)^{1/2}} \tag{1.1}$$

where P is the vapor pressure, m is the mass of the vapor atom, and k is the Boltzmann constant.

Equation (1.1) for an ideal gas is easy to derive using several standard assumptions from elementary statistical mechanics.

In an ideal gas, the translational motion of the gas particles can be decomposed into motion along each of the three Cartesian coordinates, x, y, and z. We are at liberty to align the axis system so that the x direction is perpendicular to the planar surface of area A, which will be struck by the gas particles. *Motion* in the y and z directions does not carry a particle closer to the surface and can be ignored. Let $\alpha(\dot{x})\,d\dot{x}$ be the number of particles with velocities between $\dot{x}$ and $\dot{x} + d\dot{x}$ that strike the surface per unit time. In one second any particle traveling toward the surface that is at a distance x' where $x' \leq x = \int_0^1 \dot{x}dt$ will strike the surface. Therefore, the number of particles with velocity $\dot{x}$ that will strike the surface of area A in one second is equal to the number of $\dot{x}$ particles within the volume $A \cdot x$. Letting $p(\dot{x})$ be the number density of particles with velocity $\dot{x}$, we may write

$$\alpha = \int_0^\infty \alpha(\dot{x})\,d\dot{x} = \int_0^\infty \dot{x}p(\dot{x})\,d\dot{x}$$
$$= \int_0^\infty \frac{\dot{x}N}{V}P(\dot{x})\,d\dot{x} \tag{1.2}$$

where

$$P(\dot{x}) = \frac{\Omega(\dot{x})\exp(-m\dot{x}^2/2kt)}{\int \Omega(\dot{x})\exp(-m\dot{x}^2/2kt)\,d(1/2m\dot{x}^2)} \tag{1.3}$$

where V is the volume and N is the number of particles in the volume. Equation (1.3) is a form of the Gibbs distribution law:

$$p(\varepsilon) = \frac{\Omega(\varepsilon)\exp(-\varepsilon/kt)}{\int \Omega(\varepsilon)\exp(-\varepsilon/kt)\,de} \tag{1.4}$$

where $\Omega(\varepsilon)$ is the degeneracy of the energy state ε and is equal to the volume in phase space divided by h^{3N}.[14] For the uninitiated, this means that if we write the energy in terms of momentum and integrate over the volume of the container as well as over the appropriate momenta, the degeneracy $\Omega(\varepsilon)$ is automatically incorporated.

So Eq. (1.2) may be rewritten

$$\alpha = \frac{\int \cdots \int_{0}^{\infty} P_x N/mV \exp(-P_x^2/2mkT)\, dP_x/m \, dx \, dP_y \, dy \, dP_z \, dz}{\int\int \cdots \int_{-\infty}^{+\infty} \exp(-P_x^2/2mkT)\, dP_x/m \, dx \, dP_y \, dy \, dP_z \, dz}$$

The indefinite integrals in the numerator and denominator can be written indefinite because they cancel in any case.

In the numerator, P_x is integrated from 0 to ∞ instead of $-\infty$ to $+\infty$ because only particles moving toward the surface are to be counted.

Using $\int_0^\infty xe^{-ax^2} = 1/(2a)$ and $\int_{-\infty}^{+\infty} e^{-ax^2} = 1/2\sqrt{\pi/a}$, we see that $\alpha = P(2\pi mkT)^{-1/2}$, as shown in Eq. (1.1). From the expression for the chemical potential μ of an ideal gas,[6]

$$\mu = -kT \ln\left[\left(\frac{2\pi mkT}{h^2}\right)^{3/2} \frac{V}{N} f_g(T)\right]$$

we obtain

$$\alpha = kT \frac{(2\pi mkT)}{h^3} e^{\mu/kT} f_g(T) \tag{1.5}$$

where $f_g(T)$ is the internal partition function for atoms in the gas phase.

Fowler[7] derived an equation using statistical mechanics for an ideal lattice gas system on a two-dimensional, uniform lattice with no interaction between nearest neighbor adsorbed atoms. Fowler's treatment gives rise to an interesting expression for the rate of evaporation of gas atoms from the lattice surface:

$$\nu = k_0 e^{-E/kT} \tag{1.6}$$

where

$$k_0 = \frac{(2\pi mkT)kT}{h^3} \frac{f_g(T)}{f_a(T)}$$

E is the binding energy per particle per site, and $f_a(T)$ is the internal partition function for a particle in the adsorbed phase and is assumed to be independent of the local environment of the adsorbed particle. (This is not necessarily realistic because f_a depends on the number of degrees of freedom of a particle, which varies with the number of nearest neighbors.)

Equations (1.5) and (1.6) are quite similar. In fact, using these equations to form the ratio of the rate at which particles strike the surface to the rate at which they evaporate yields

$$\frac{\alpha}{\nu} = e^{(E+\mu)/kT} \tag{1.7}$$

Note that we have set $f_a(T) = 1$ in deriving Eq. (1.7).

In computer simulation of the nonequilibrium behavior of a system, the relative rates for the various processes must be known. Equation (1.7) expresses the needed relative rate for adsorption and evaporation. We further assume that the nearest neighbor interactions and next nearest neighbor interactions affect only the energy of the interaction E and not the form of Eq. (1.7). That is,

$$E = -\varepsilon - \eta w - \eta' w' \tag{1.8}$$

where ε is the binding energy of a site, η and η' are the number of nearest neighbor and next nearest neighbor atoms, respectively, and w and w' are the nearest neighbor and next nearest neighbor lateral interaction energies, respectively. The minus sign precedes ε, w, w' so that a negative ε or (w, w') is an attractive bond and a positive ε or (w, w') is repulsive. Eq. (1.7) becomes

$$\frac{\nu(\eta, \eta')}{\alpha} = \exp\left(\frac{\varepsilon + \eta w + \eta' w'}{kT}\right) \tag{1.9}$$

For future reference the following quantities are defined:

$$P/P^* \equiv \exp\left(\frac{\mu - \varepsilon}{kT}\right) \tag{1.10a}$$

$$C \equiv \exp\left(\frac{-w}{kT}\right) \tag{1.10b}$$

$$C' \equiv \exp\left(\frac{-w'}{kT}\right) \tag{1.10c}$$

For a square lattice with van der Waal–London interactions, we find $C' = C^{1/8}$.

II. THE COMPUTER SIMULATION MODEL

A. Boundary Conditions

In the Vapor Deposition Simulation (VDS) programs, the surfaces on which vapor atoms are adsorbed are two-dimensional square lattices. These lattices might, e.g., represent the (100) planes in a cubic closest packed crystal. Periodic boundary conditions are used, i.e., for an $N \times N$ lattice, $X_1 = X_{N+1}$ and $Y_1 = Y_{N+1}$.

B. Evaporation, Migration, and Nearest Neighbor Effects

We assume that the binding energy of an adsorbed atom to the substrate is related only to the number and type of atoms it "touches." We assume that there exist only two types of atoms: the crystalline surface atoms and the vapor atoms. The atoms are the same size so that the deposited atoms will retain the crystal packing.

The crystalline surface atoms are fixed once their positions are specified at the beginning of the problem. We will denote these atoms by "B." The atoms of the vapor phase may condense on the lattice, evaporate, or migrate. We will denote these atoms by "W." E_{BW} denotes the binding energy for two touching W atoms. The total binding energy equals the sum of the binding energies of the single bonds.

1. Evaporation

The evaporation of an atom requires that the bonds between it and its neighbors be broken. The evaporation rate per atom ν_e is a function of the binding energy E; $\nu_e \propto \exp(E/kT)$. Examples of the binding energy for some evaporation configurations are presented in Figure 1a, b, and c. Actually, for evaporation we also consider the effect of next nearest neighbor interactions, which is not accounted for in this procedure of counting bonds through the packing configuration. However, the usefulness of this procedure becomes obvious in operationally determining the activation free energy for migration. For migration next nearest neighbor interactions are neglected.

2. Migration

The migration of an atom requires that some bonds, but not all, be broken in order to go to another site. This depends on the particular geometry for the migration. A particular example is given in Figure 1d.

If the decision about which bonds are broken or not seems artificial, it should be remembered that equilibrium properties are not altered by migration rates. This also makes experimental verification of migration rate very difficult. All that we hope to gain from the present model is a "feel" for the effects of migration on the rate at which systems come to equilibrium.

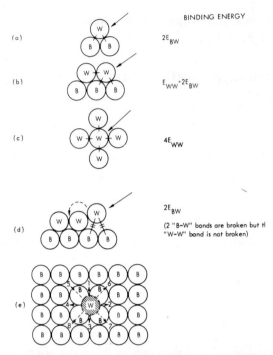

Fig. 1. The binding energy for different geometrical configurations and transitions. (e) depicts the sites to which
the W adatom may migrate for monolayer transitions.

In migration we do not consider the effect of next nearest neighbor
interactions. In order to illustrate our procedure, we present all ε_m for
monolayer migration illustrated in Figure 1e. ε_m is used to denote the energy
barrier to migration.

Case 1: The migration of the W adatom to

  a. positions 1, 2, 3, or 4 requires 2B-W bonds to be broken, i.e.,
     $\varepsilon_m = 2E_{WB}$ (no nearest neighbor adatoms)
  b. position 5, 6, 7, or 8 requires 3B-W bonds to be broken, i.e.,
     $\varepsilon_m = 3E_{WB}$ (no nearest neighbor adatoms)

Case 2: Consider the case where one neighboring adatom is resting
in position 4. The migration of the W adatom to positions

  a. 1, 2, 3 requires the $\varepsilon_m = 2E_{WB} + 1xE_{WW}$
  b. 4 requires that $\varepsilon_m = -\infty$ (no migration)
  c. 6, 7 requires that $\varepsilon_m = 3E_{WB} + 1xE_{WW}$
  d. 5, 8 requires that $\varepsilon_m = 3E_{WB}$

Case 3: Consider the case where two neighboring adatoms are resting in positions 3 and 4. The migration of the W adatom to position(s)

a. 1, 2 requires that $\varepsilon_m = 2E_{WB} + 2E_{WW}$
b. 6 requires that $\varepsilon_m = 3E_{WB} + 2E_{WW}$
c. 5, 7 requires that $\varepsilon_m = 3E_{WB} + 1xE_{WW}$
d. 3, 4, 8 is forbidden

Case 4: Consider the case where three neighboring adatoms rest in positions 2, 3, and 4. The migration of the adatom to position(s)

a. 1 requires that $\varepsilon_m = 2E_{WB} + 3E_{WW}$
b. 5, 6 requires that $\varepsilon_m = 3E_{WB} + 2E_{WW}$
c. 2, 3, 4, 7, 8 is forbidden

Case 5: No migration if neighbors exist in position 1, 2, 3, and 4.

Case 6: Consider the case where two neighboring adatoms rest in positions 1 and 3. The migration of the adatom to positions

a. 2, 4 requires that $\varepsilon_m = 2E_{WB} + 2E_{WW}$
b. 5, 6, 7, 8 requires that $\varepsilon_m = 3E_{WB} + 1xE_{WW}$
c. 1, 3 is forbidden

Case 7: Consider the case where an adatom rests in position(s) 5, 6, 7, and/or 8. The migration to

a. 5, 6, 7, and/or 8 is forbidden
b. otherwise, it is like Cases 1–6

The following "normalized" rate expressions are used:

$$\frac{v_e}{\alpha} = \exp\left(\frac{\varepsilon + \eta w + \eta' w' - \mu}{kT}\right) \qquad \text{(evaporation)} \qquad (2.1)$$

$$\frac{v_m}{\alpha} = \exp\left(\frac{\varepsilon_m - \mu}{kT}\right) \qquad \text{(migration)} \qquad (2.2)$$

where ε_m is the energy associated with the bonds broken in order for the adatom to migrate to the neighboring site.

C. Initial Conditions

The initial coverage in the VDS programs may be set equal to zero or may be any fraction of the total number of sites, with the particles randomly distributed over the surface.

III. RANDOM NUMBERS AND SIMULATION STRATEGY

A. Monte Carlo Methods

As indicated earlier, random numbers are employed to control the dynamic system of atoms evaporating, condensing, and/or migrating on the surface of a lattice. The VSD programs are thus based on Monte Carlo methods. Monte Carlo methods[8] may be defined as those that use random numbers to solve problems in applied mathematics. Their importance rests largely on the availability of high-speed electronic computers, and so this branch of applied mathematics is still in its infancy.

B. Generation of Random Numbers

Random numbers are numbers with a rectangular distribution between two specified limits. There apparently is no number theoretic algorithm for generating random numbers. The best tables of random numbers have been prepared with the aid of physical processes that evidently have a random character.

Pseudo-random numbers with rectangular distributions acceptable for most applications can be generated in computers with 32 bit words as follows: Let R_i be a random number between 1 and 2^{31}. Then a new pseudo-random number can be generated by Eq. (3.1):

$$R_{i+1} = (aR_i + c) \bmod 2^{31} \qquad (3.1)$$

Equation (3.1) is an example of a mixed congruential random number generator. A discussion of congruential random number generators is given by Marsaglia and MacLaren.[11] Most computer-generated random numbers are created in this fashion, although the mod 2^{31} may change, depending on the machine word length. Also, either a or c may be set to zero, producing additive or multiplicative congruential generators.

The inspiration for Eq. (3.1) apparently springs from a suggestion made by Metropolis and von Neumann that random numbers could be generated by squaring a number of, say, length M and then taking M digits (or bits) from the central portion of the squared number. The new number gained from the central portion of its squared predecessor is taken as a new random number.

When Eq. (3.1) is implemented on a computer, the operation of finding the modulus is performed by shifting bits off the end of the word.

IBM supplies a subroutine called RANDU of the multiplicative congruential type in their Scientific Subroutine Package. It is worth examining this subroutine to illustrate Eq. (3.1).

```
  SUBROUTINE RANDU (IX, IY, YFL)
  IY = IX*65539
  IF(IY) 5, 6, 6
5 IY = IY + 2147483647 + 1
6 YFL = IY
  YFL = YFL*4656613E − 9
  RETURN
  END
```

The integer 65539 plays the role of a in Eq. (3.1). It shifts out approximately the upper half of the word containing IX, the initial random number. If the sign bit of the new random number IY is a one, then $2^{31} + 1$ is added to IY to produce a positive number. Finally, the random number is stored in YFL and divided by 2^{31} to scale it to the range (0, 1.0).

The random numbers produced by RANDU have long chain patterns that would bias results based on a great many random numbers from RANDU. Better random numbers are obtainable by using a random switching table in which the random number supplied to RANDU in selected from a table of random numbers, which is, in turn, indexed by the previous random number. This approach can be further "randomized" by more switching tables but at the cost of slower execution. However, very good random numbers are obtainable in this fashion.

Another method of avoiding undesired correlations between processes is to use several different random number chains, in particular, the X coordinate and Y coordinate of a randomly chosen point on an XY plane are each generated from independent random number chains in the VDS program.

C. Gaussian Distribution Generated by Random Numbers

A random variable distributed according to a Gaussian distribution is easy to generate with random numbers. The trick is to take advantage of the central limit theorem, which says, in this case, that the sum of n random variables distributed uniformly about zero approaches a Gaussian distribution with mean zero as $n \to \infty$. In practice, an acceptable Gaussian distribution is usually obtained when n is larger than 10 or 12.

The variance σ^2 of the sum of n random numbers, each of which is uniformly distributed between -0.5 and $+0.5$, is calculated as follows:

$$\sigma^2 = \int_0^{0.1} \cdots \int_0^1 \left(\sum_{i=1}^n (\mathrm{YFL}_i - 0.5) \right)^2 d\mathrm{YFL}_1\, d\mathrm{YFL}_2 \cdots d\mathrm{YFL}_n$$

A change of variables, setting $R_i = \mathrm{YFL}_i - 0.5$, leads to

$$\sigma^2 = \int_{-0.5}^{+0.5} \cdots \int_{-0.5}^{+0.5} \left(\sum_i R_i \right)^2 dR_1 \cdots dR_n$$

but

$$\int_{-\frac{1}{2}}^{+\frac{1}{2}} R_i\, dR_i = 0 \quad \text{and} \quad \int_{-\frac{1}{2}}^{+\frac{1}{2}} dR_i = 1$$

Therefore,

$$\sigma^2 = \int_{-0.5}^{+0.5} \cdots \int_{-0.5}^{+0.5} \sum_i (R_i)^2\, dR_1 \cdots dR_n = \sum_i \int_{-0.5}^{+0.5} R_i^2\, dR_i = \frac{n}{12}$$

So if we choose $n = 12$, the variance is unity. SUBROUTINE GAUSS, an IBM Scientific Subroutine, uses these tricks:

```
          SUBROUTINE GAUSS (IX, S, AM, V)
          A = 0.0
          DO 50 I = 1, 12
          CALL RANDU (IX, IY, YFL)
          IX = IY
     50   A = A + YFL
          V = (A − 6.0)*S + AM
          RETURN
          END
```

where S is the standard deviation, AM is the mean, and V is the resultant random variable selected from a Gaussian distribution with mean = AM and variance = S.

There is another method of producing random variables according to a Gaussian distribution. This method is general and can be used to produce random variables according to any given distribution function. The technique is very similar to the Monte Carlo method for integrating a function. A random number is generated, scaled, and tested to see if it lies above or below the curve or plane determined by the function:

```
      SUBROUTINE GENORM (IX, S, AM, V)
      PI = 3.14159
  1   CALL RANDU (IX, IY, V)
      IX = IY
      V = 4*S*(V − 0.5)
      CALL RANDU (IX, IY, Y)
      IX = IY
      F = 1.0/(EXP(((V/S)**2)/2)*S*SQRT(2*PI))
      IF(F − Y) 1, 2, 2
  2   V = V + AM
      RETURN
      END
```

Since integration is basically a means for finding the area, volume, or hypervolume under a curve, plane, or hyperplane, the technique used in SUBROUTINE GENORM is also used to perform integrations. Random numbers are used to fill a known volume V_1 with a known number n_1 of randomly placed points. The number n_2 of points that lie under the surface of the function being integrated are counted. The value of the integral V_2 is given by

$$V_2 = \frac{n_1 V_1}{n_2}$$

Random numbers perform other functions in simulation problems. They often choose when, where, and what event is to happen next. This is illustrated by the man, the two women and the commuter train problem.

A young Stanford man knows two women whom he likes equally well, one in San Francisco and one in San Jose. He knows that commuter trains run one per hour to the two cities. He therefore acts on the assumption that if he arrives at the commuter train depot at random times and boards the first train to arrive then he will, on the average, visit both girls equally often. However, the young man soon discovers that this method carries him to San Francisco twice as often as it does to San Jose.

What is the fallacy? Although the young man arrives at the station at random times, the trains do not—they run on a fixed schedule. Thus if the San Francisco trains leave on the hour and the San Jose trains happen to leave at 20 minutes after the hour, the young man will be twice as likely to arrive during the period before the San Francisco-bound train leaves.

This problem illustrates how a random number—the young man's arrival time at the depot—selects a process: the trip to San Francisco or San Jose. This is the method used to select evaporation, condensation, or migration of molecules in a lattice–gas system: a random number is generated and the process associated with the magnitude of the number is selected.

D. Use of Random Numbers to Select Dynamic Processes—Simulation Strategy

Our application of the Monte Carlo method to vapor deposition was inspired by Richard Gordon's[9] original paper on the computer simulation of adsorption isotherms of lattice gases.

Using random numbers to control the various movements of a dynamic system of atoms condensing, evaporating, and/or migrating on the surface of a lattice deserves special attention.

All motion must be characterized in terms of motion in discrete directions and for discrete distances. For the VDS programs, motion is restricted to eight directions on the lattice surface and two directions perpendicular to the lattice surface (up and down). Care must be taken to preserve the relative magnitudes of these rates as found in nature. In particular, if a uniformly distributed random number between zero and one is used to select the direction of migration and/or evaporation of a particle, then the sum of the rates of the nine possible processes must sum to one or less. But when normalizing the eight rates of migration and the rate of evaporation to ensure that their sum is one or less, the rate for condensation must also be normalized by the same factor to preserve proper relative magnitudes.

In the VDS programs, the following technique is used to select various processes: A site is chosen by two random numbers representing the X and Y coordinates.

Case A: The rate of condensation is less than the sum of the rates for migration and evaporation. If the randomly chosen site is empty, a random number between zero and one that falls inside or outside a "bin" determines whether a particle should condense on that site, i.e., there is a "null-event" bin and a condensation bin. The bin is actually a line segment L, $0 \leq L \leq 1$. If L is proportional to the rate of condensation, a random number falling on a point contained in L determines a condensation event and a particle is deposited on the site. If a randomly chosen site is occupied, the particle at the site may migrate in eight directions or evaporate in one direction. To choose between the various possibilities, a random number is produced and tested to see in which bin it lies. The size of the bin is proportional to the rates of the processes, migration and evaporation. The process associated with the bin containing the random number is executed.

Case B: The rate of condensation is greater than the sum of the rates for migration and evaporation. For this case, the null-event bin must be included among the possible processes to be considered for an occupied site, i.e., the null-event bin is not part of the possible processes to be considered when an empty site is chosen. If a randomly chosen site is empty, a particle is placed on it. The VDS programs automatically place the null-event bin where it belongs.

IV. REAL AND SIMULATED TIME

Time as used by the laboratory experimentalist is a continuous well-defined variable, whereas time in Monte Carlo simulations is quantized and perhaps of variable duration when measured by the laboratory clock. Can the real time (laboratory time) rate of a physical process be related to the simulation rate? The answer is "yes" in the limit of large times or a large number of simulated events. This may not be obvious, so we will now derive a relationship between the real time rate of events occurring at random times (e.g., vapor deposition or migration) and the simulated rate of events.[17]

The Poisson density function (Feller,[15] page 146) shown in Eq. 4.1 describes a process that occurs at random times with the mean number of events in time t being λt. Let λ be the rate at which some physical event occurs in the laboratory, say, the rate at which particles evaporate from lattice sites. The probability that K particles evaporate in time t is given in Eq. (4.1):

$$p(K, \lambda t) = \frac{e^{-\lambda t}(\lambda t)^K}{K!} \tag{4.1}$$

In the simulated system, the probability of K particles evaporating in T event time units is binomial (Feller,[15] page 136):

$$b[K, T, P] = \binom{T}{K} P^K (1 - P)^{T-K} \tag{4.2}$$

It is well known that the Poisson distribution uniformly approximates the binomial distribution for the proper choices of variables. Following Feller, we see that setting TP equal to λt minimizes the difference between the binomial and Poisson distributions. For large K, when K is close to λt, as it will be under the conditions relevant to this discussion, the two distributions become nearly identical with $\lambda t = TP$.

Observe that we have just written an explicit expression relating the real time t (measured in the laboratory) to an event time T (measured during the simulation):

$$t = \frac{TP}{\lambda} \qquad (4.3)$$

In vapor deposition, e.g., T might be the number of sites chosen on the lattice surface, P would then be the product of the probability that the chosen site is empty (which would be a function of the lattice coverage) and of the probability that an empty site receives an adatom (which is a fixed parameter), and λ would be the mean rate of deposition measured in the laboratory per site (not per unit area).

If it were of interest to calculate the real time lag for the growth of an initially empty lattice to its equilibrium coverage, it would be necessary to measure the functional dependence of θ, the fractional lattice coverage, on T. This is because P is a function of θ. When $P(\theta)$ has been converted to $P(T)$ a numerical integration of the derivative form of (4.3) could be performed to yield the desired lapse time.

V. THE VDS PROGRAMS

A flow diagram describing the calculational procedure of the VDS programs is presented in Figure 2. Three different programs make up the VDS program package. They are called HPLOT, MIGRATE, and BETH3D.

HPLOT, which is short for Honig (adsorption isotherm) Plot, is a program that produces a Honig adsorption isotherm by simulating vapor phase deposition onto a square lattice. Section IB describes the Honig model.

The purpose of MIGRATE is to simulate the vapor phase deposition of atoms on a lattice surface where the adsorbed atoms may migrate. The lattice consists of squares, and the adsorption is restricted to monolayer. Migration occurs as described earlier. Only nearest neighbor interactions are considered.

BETH3D (BET, Honig, 3 Dimensional, adsorption isotherms) is the same as HPLOT except that the number of particles allowed to be absorbed on a site can be greater than one and the stacking is cubical with higher cubes resting directly over lower neighbors. (Therefore, the binding energy

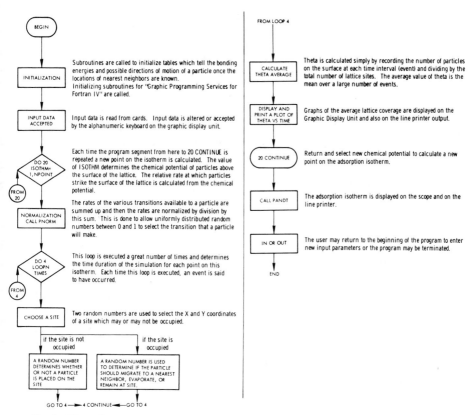

Fig. 2. Flow diagram describing the calculational procedure of the Vapor Deposition Simulation programs.

between first adsorbed cubes and the lattice is E_{BW} instead of $4E_{BW}$). Only nearest neighbor interactions are considered. The well-known BET isotherm[6] may be simulated by excluding lateral interactions.

The IBM 2250 Diplay Unit and the subroutine package called Graphic Programming Services for FORTRAN IV are used. The programs may receive input data from the scope's alphameric keyboard. Results of simulations are summarized in plots of adsorption isotherms and also in plots of the relative number of particles on the lattice as a function of time. A "snapshot" of the lattice also is given at specified time intervals that shows the locations of the particles on the lattice surface. Pictures and graphs displayed on the scope also are recorded on the line printer output so that a permanent record of results is maintained. Graphs displayed on the scope may be photographed for permanent record.

The IBM S/360 Model 50 running time to generate 100,000 elementary events in a particular simulation is approximately 2.5 minutes.

VI. THE COMPUTER SIMULATION RESULTS

In Figures 3, 4, and 5 the results of some vapor deposition simulations are summarized in plots of coverage versus time and adsorption isotherms. The coverage scale is normalized so that the average coverage (theta average on the plots) lies approximately midway on the graph. This normalization was chosen in order to facilitate our analysis of a particular computer run via the 2250 Display Unit. Time is measured in terms of "events" where an event is the selection of a lattice site prior to considering changing that site by an adsorption, migration, or evaporation.

The frequency of "events" is related to the frequency with which particles are adsorbed, which in turn is related to the frequency of vapor atoms striking the surface. Thus the kinetic theory of gases through Eq. (1.1) provides a relation between an "event" and real time measure as explained in Section IIID.

It should be noted that an "event" in a system with migration (e.g., MIGRATE) will correspond to a shorter time interval than an "event" in a system with only evaporation and condensation (HPLOT or BETH3D).

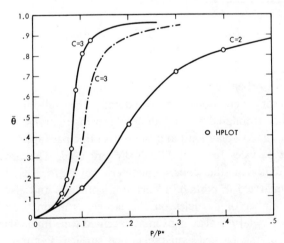

Fig. 3. Comparison of the experimental values from HPLOT with the theoretical adsorption isotherm of Honig (solid curves for $C = 2,3$. The Fowler–Guggenheim isotherm[10] is the dot–dash curve.

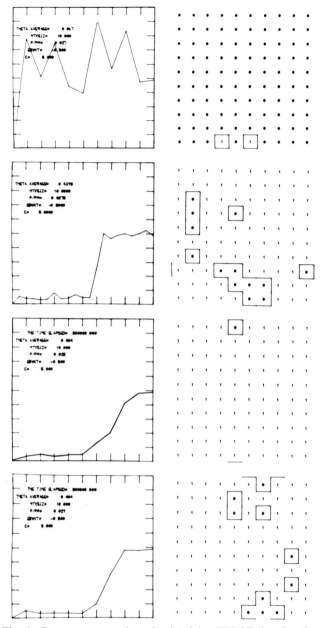

Fig. 4. Coverage versus time simulated by HPLOT for $C = 5$ and different P/P^*. "Snapshots" of the lattice are presented at the end of the time intervals to show the locations of atoms on the lattice surface. Time is measured in terms of the number of events that occurred up to that given time. The elapsed time for the first two experiments is 10^6 events. The theta average is essentially an average over the entire time region.

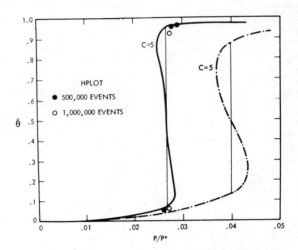

Fig. 5. Comparison of the experimental values from HPLOT with the theoretical adsorption isotherm of Honig (solid curve) for $C = 5$. The Fowler–Guggenheim isotherm[10] is the dot–dash curve.

This is easily seen by noting that the frequency with which "events" result in the selection of a condensation transition (vapor atom → adatom) is reduced when an event must sometimes choose migration as well as evaporation or condensation. The "zero" on the snapshots designates an unoccupied lattice site, and a "number" designates pile height of condensed atoms on that particular lattice site. For HPLOT and MIGRATE the pile height always will be one since they are monolayer programs.

Figure 3 demonstrates that there exists very good agreement between the experimental values from HPLOT and the theoretical isotherm of Honig.[3] MIGRATE includes the effect of migration of adatoms (a process not accounted for by Honig's theory), but it does not have next nearest neighbor interactions. $\bar{\theta}$ (MIGRATE) equals $\bar{\theta}$ obtained from simulations where migration and next nearest neighbor interactions are excluded, i.e.,

Table I. MIGRATE Adsorption Isotherm for Selected Values of P/P^*

P/P^*	0.1	0.3	0.4	0.5
$\bar{\theta}$	0.13	0.59	0.72	0.80

using HPLOT with no next nearest neighbor interactions. This feature is consistent with equilibrium statistical mechanics. The results of the MIGRATE simulations are presented in Table I.

In Figure 3 the coverage versus time is presented from HPLOT for $C = 3$ and $P/P^* = 0.07, 0.08, 0.09, 0.12$. Figure 3 demonstrates that the experimental points of HPLOT and Honig's theoretical adsorption isotherm are in excellent agreement.

Figure 4 presents the coverage versus time from HPLOT, $C = 5$ and $P/P^* = 0.027, 0.0275, 0.028, 0.029$. Each point θ in the plot represents an average theta calculated over the preceding number of events defined by the time interval of the plot. The theta average $\bar{\theta}$ printed on the plot is an average theta calculated from the time that the curve first has a negative inflection to the end time of the simulation. $\bar{\theta}$ is then used to generate the adsorption isotherm, i.e., $\bar{\theta}$ versus P/P^*. The theoretical adsorption isotherm predicts a phase transition for $P/P^* \simeq 0.0268$ (Figure 5). The HPLOT simulation indicates a transition for $0.027 < P/P^* < 0.0275$, in good agreement with Honig's calculations. We note from Figure 4 that for $P/P^* = 0.0275$, θ stayed at very low coverage until time $\simeq 5.5 * 10^5$ events and then proceeded to grow to $\bar{\theta} = 0.92$ over a time interval of 10^5 events. For higher P/P^*, i.e., 0.028 and 0.029, it took $\sim 2.5 * 10^5$ events before growth began. Possibly the nucleation rate at $P/P^* = 0.027$ is such that we would have to look at more than 10^6 events in order to observe the phase transition. However, limits on the computational time available prevented us from extending the simulation.

Finally, in this study we choose to simulate the BET isotherm (Hill,[6] page 134) using the BETH3D program. The BET isotherm equation is

$$\theta(1 - G)\left[\frac{1}{S_C} + G\left(1 - \frac{1}{S_C}\right)\right] \tag{6.1}$$

where

$$\theta \equiv \frac{\text{total number of particles adsorbed}}{\text{total number of sites available}} = \frac{A}{B}$$

$$S_C = \exp[(E_{WW} - E_{BW})/kT]$$

and

$$G \equiv P/P^*$$

Table II compares the predictions of BET theory (Eq. (6.1)) with the experimental values from BETH3D. We note good agreement.

Table II. BET Adsorption Isotherm for Selected Values of $P/P*$ [a]

$P/P*$	0.1	0.2	0.3	0.4	0.5	0.6	0.7	0.8
ABT	0.58	0.89	1.16	1.45	1.82	2.34	3.20	4.87
ABS	0.57	0.87	1.15	1.42	1.93	2.36	3.16	4.62

[a] $ABT = A/B$ (theoretical) and $ABS = A/B$ (simulated). A/B (theoretical) is defined by Eq. (6.1) in the text.

Honig's theoretical results are difficult to verify experimentally in the laboratory and so have remained without experimental corroboration until now. The results of our simulations indicate that Honig's treatment is quite accurate and is a substantial improvement over previous treatments. This agreement also serves to build confidence in the use of Monte Carlo methods in simulating the molecular dynamics of vapor deposition.

The principal application was to systems that permit a comparison with Honig's work, although the simulation methods are easily applied to a variety of other problems.

An overall objective of the Vapor Deposition Simulation Programs[1] is to provide materials scientists with a flexible subroutine-oriented computer program package which, with minor alterations, should encompass a wide range of problems: condensation and/or evaporation and/or migration on 3-D or 2-D crystal surfaces including the effects of several different lattice-adatom binding energies, lattice defects, first and second nearest neighbor binding energies, and different lattice–adatom packing arrangements. Although not all of these features are incorporated in the present versions of the programs, the missing features can be inserted with relative ease.

References

1. F. F. Abraham and G. M. White, Computer simulation of vapor deposition on two-dimensional lattices, *IBM Palo Alto Scientific Center Report No. 320-3252* (1969).
2. E. A. Flood, ed., *The Solid Gas Interface*, Marcel Dekker, Inc., New York, 1967, Vol. 1.
3. J. M. Honig, Adsorption theory from the viewpoint of order–disorder theory, p. 371 of Reference 2 above.
4. J. P. Hirth and G. M. Pound, *Condensation and Evaporation*, Pergamon Press, London, 1963.

5. J. Hijmans and J. deBoer, An approximation method for order-disorder problems I, *Physica* **21**, 471–484 (1955).
6. T. L. Hill, *Statistical Thermodynamics*, Addison-Wesley Publishing Co., Inc., Reading, Mass., 1960, p. 80.
7. R. H. Fowler, A statistical derivation of Langmuir's adsorption isotherm, *Proc. Cambridge Phil. Soc.* **31**, 260–264 (1935).
8. J. M. Hammersly and D. C. Handscomb, *Monte Carlo Methods*, Methuen and Co., Ltd., London, 1964.
9. R. Gordon, Adsorption isotherms of lattice gases by computer simulation, *J. Chem. Phys.* **48**, 1408–1409 (1968). Gordon gives earlier references that describe some previous work on computer simulation of irreversible vapor deposition.
10. R. H. Fowler and E. A. Guggenheim, *Statistical Thermodynamics*, Cambridge University Press, New York, 1939.
11. G. Marsaglia and M. D. MacLaren, Uniform random number generators, *J.A.C.M.* **12**, 83–89, 1965.
12. A. J. W. Moore, Nucleation of solids from the vapor phase, *J. Austr. Inst. Metals*, **11**, 220–226 (1966).
13. G. H. Gilmer and P. Bennema, *J. Appl. Phys.* **43**, 1347 (1972).
14. F. Reif, *Fundamentals of Statistical and Thermal Physics*, McGraw-Hill Book Co., 1965, p. 64.
15. W. Feller, *An Introduction to Probability Theory and Its Applications*, John Wiley and Sons, Inc., New York, 1957.
16. F. F. Abraham and F. G. White, Computer simulation of vapor deposition on two-dimensional lattices, *J. Appl. Phys.* **41**, 1841 (1970).
17. G. M. White and F. F. Abraham. Simulation time versus real time in computer simulation of vapor deposition, *J. Appl. Phys.* **41**, 5348–5350 (1970).

APPENDIX:
Fortran Code

The main program segment listing shown below, together with the listed subroutines, form the heart of the VSD program.

When the main program segment shown below is entered, a site (*IX*, *IY*) has been chosen. The site also has been tested and found to be occupied by an adatom. The task of the program segment is to determine if a migration or evaporation should occur and, if migration is chosen, to which site the adatom should go.

The first thing the program segment does is determine the number of first and second nearest neighbors. It also determines the "pattern" of nearest neighbor adatoms (which will be one of the six basic types "rotated" through 0°, 90°, 180°, or 270°).

(The sites that are available to receive a migrating adatom are known to the program through the classification of the "pattern" of nearest neighbors. This is done by SETPAT and LOCGEN. A table lookup then can be performed to locate the site receiving the migrating adatom by simply "rotating" the coordinates of the site chosen by the random number. This is done by ROTATE.)

The program next enters code segment Binscan, BSAN for short, which "scans bins" to see which bin contains a newly created random number. The "bins" are labeled evaporation, null event, and a maximum of eight different directions of migration. The size of the bins are determined by the number of nearest neighbors. On the basis of this information, an evaporation or migration or null event is chosen. These bins do not distinguish between two energetically identical but geometrically dissimilar configurations of nearest neighbors. But the pattern recognition function SETPAT does distinguish geometrically dissimilar patterns and returns a value IROTATE, which tells how far a given pattern must be rotated to match another. IROTATE is then used in LOC, which finally determines which site receives the migrating adatom.

In order that this segment of code be better understood, the following subroutines should be read: PNORM, ENGEN, MIRATE, LOCGEN, ROTATE, SETPAT, MINORM, and NORM.

```
      SUBROUTINE PNORM(EBWKT,EWWKT,EWWKTP,UKT,P,P2,PU,R,SUM)
      DIMENSION  P2(1)
      DIMENSION  P(1)
      DIMENSION  R(6,1)
C     PNORM GENERATES   P AND P2 VECTORS, CALLS ON MIRATE-(MIGRATION RATES)
C     -- ,CALLS ON MINORM,WHICH NORMALIZES ALL THE VECTORS, AND THEN
C     DIVIDES OUT THE ENERGY OF INTERACTION BETWEEN THE LATTICE SURFACE
C     AND THE FIRST LAYER OF ADSORBED PARTICLES.
C     THE VALUE OF EBWKT SUPPLIED TO THIS PROGRAM MUST BE THE LARGEST
C     VALUE OF THE BLUE-WHITE BINDING ENERGIES IF MORE THAN ONE BINDING
C     ENERGY IS INVOLVED (AS IN THE CASE OF RANDOM BINDING ENERGIES.)
      PU=EXP(UKT)
      DO 7 I=1,4
      AI=I
    7 P2(I+1)=EXP(AI*EWWKTP)
      P2(1)=1.0
      P(1)=EXP(4*EBWKT)
      P(2)=P(1)*EXP(EWWKT)
      P(3)=P(2)*EXP(EWWKT)
      P(4)=P(3)*EXP(EWWKT)
      P(5)=P(4)*EXP(EWWKT)
      CALL MIRATE(R,EBWKT,EWWKT)
      CALL MINORM(R,P,PU,SUM)
      DO 3 I=1,5
    3 P(I)=P(I)*EXP(-4.0*EBWKT)
      RETURN
      END



      SUBROUTINE  ENGEN(MTXSIZ,EBWKT,RATIO)
      COMMON  SITE(50,50),ENERGY(50,50)
      INTEGER SITE
C     THIS SUBROUTINE GENERATES THE LATTICE-PARTICLE BINDING ENERGIES.
C     EFFECTS OF  'LATTICE DEFECTS' IN ENERGY ARE INCORPORATED HERE.
C     A DIAGONAL LINE OF SITES WITH BINDING ENERGIES,EBWKTL, DETERMINED
C     BY RATIO IS PUT IN HERE.
      EBWKTL=EBWKT*RATIO
      X1=EXP(4*EBWKTL)
      X=EXP(4*EBWKT)
      DO 1 I=1,MTXSIZ
      DO 1 J=1,MTXSIZ
    1 ENERGY(I,J)=X
      DO 2 I=1,MTXSIZ
    2 ENERGY(I,I)=X1
      RETURN
      END



      SUBROUTINE  NORM(EBWKT,EWWKT,EWWKTP,UKT,P,P2,PU,PMAX)
      DIMENSION  P2(1)
      DIMENSION  P(1)
C     THIS  SUBPROGRAM CALCULATES P(I)  AND NORMALIZES  THE VECTOR
C     COMPONENTS SO THAT THE GREATEST  ONE IS EQUAL TO 1.
C     THE VALUE OF EBWKT SUPPLIED TO THIS PROGRAM MUST BE THE LARGEST
      PU=EXP(UKT)
      DO 7 I=1,4
      AI=I
    7 P2(I+1)=EXP(AI*EWWKTP)
      P2(1)=1.0
      P(1)=EXP(4*EBWKT)
      P(2)=P(1)*EXP(EWWKT)
      P(3)=P(2)*EXP(EWWKT)
      P(4)=P(3)*EXP(EWWKT)
      P(5)=P(4)*EXP(EWWKT)
      PMAX=PU
      DO 2 I=1,5
      IF(P(I)-PMAX) 2,2,1
    1 PMAX=P(I)
    2 CONTINUE
      IF(EWWKTP) 9,9,4
    4 DO 6 I=1,5
      DO 6 J=2,5
      IF(P(I)*P2(J)-PMAX)  6,6,5
    5 PMAX=P(I)*P2(J)
    6 CONTINUE
    9 CONTINUE
      PU=PU/PMAX
      DO 3 I=1,5
    3 P(I)=P(I)*(EXP(-4.0*EBWKT))/PMAX
      RETURN
      END
```

```
      SUBROUTINE  MIRATE(R,EBWKT,EWWKT)
C     THIS SUBROUTINE  CALCULATES   THE TRANSITION RATES FOR MIGRATION
C     WITH  NN  NEAREST  NEIGHBORS  WHERE  NN  RANGES FROM ZERO  TO FOUR.
C     THE BONDING ENERGY  EFFECTS OF SECOND NEAREST NEIGHBORS ON TRANSITION
C     RATES ARE NOT INCLUDED.    THE TRANSITION RATES CALCULATED HERE ARE
C     CONNECTED WITH THE RATES FOR EVAPORATION AND CONDENSATION IN
C     THE SUBROUTINE MINORM.
C     R(I,J) IS THE RELATIVE RATE OF MIGRATION TO LOCATION J FOR A PARTICLE
C     IN STATE I, WHICH IS ONE OF THE SIX POSSIBLE STATES DESCRIBED IN
C     ABRAHAMS  NOTES.
      DIMENSION  R(6,8)
C     STATE ONE  ---  NO NEAREST NEIGHBORS---  R(1,J) IS CALCULATED.
      DO 1 I=1,4
    1 R(1,I)=EXP(2.0*EBWKT)
      DO 2 I=5,8
    2 R(1,I)=EXP(3.0*EBWKT)
C     STATE  TWO  --- ONE NEAREST  NEIGHBOR --- R(2,J)
      DO 3 I=1,3
    3 R(2,I)=EXP(2.0*EBWKT+EWWKT)
      R(2,6)=EXP(3.0*EBWKT+EWWKT)
      R(2,7)=EXP(3.0*EBWKT+EWWKT)
      R(2,5)=EXP(3.0*EBWKT)
      R(2,8)=EXP(3.0*EBWKT)
      R(2,4)=0
C     TWO  NEAREST NEIGHBORS---R(3,I)
      R(3,1)=EXP(2.0*EBWKT+2.0*EWWKT)
      R(3,2)=EXP(2.0*EBWKT+2.0*EWWKT)
      R(3,6)=EXP(3.0*EBWKT+2.0*EWWKT)
      R(3,5)=EXP(3.0*EBWKT+1.0*EWWKT)
      R(3,7)=EXP(3.0*EBWKT+1.0*EWWKT)
      R(3,3)=0
      R(3,4)=0
      R(3,8)=0
C     THREE NEAREST NEIGHBORS ---- R(4,J)
      DO 5 I=1,8
    5 R(4,I)=0
      R(4,1)=EXP(2.0*EBWKT+3.0*EWWKT)
      R(4,5)=EXP(3.0*EBWKT+2.0*EWWKT)
      R(4,6)=EXP(3.0*EBWKT+2.0*EWWKT)
C     FOUR  NEAREST NEIGHBORS---R(5,J)
      DO 6 I=1,8
    6 R(5,I)=0
C     TWO  NEAREST NEIGHBORS --- R(6,J)
      R(6,2)=EXP(2.0*EBWKT+2.0*EWWKT)
      R(6,4)=EXP(2.0*EBWKT+2.0*EWWKT)
      R(6,1)=0
      R(6,3)=0
      DO 7 I=5,8
    7 R(6,I)=EXP(3.0*EBWKT+EWWKT)
      RETURN
      END



      SUBROUTINE  LOCGEN(LOC)
      DIMENSION  LOC(2,4,1)
C     THE VALUES GENERATED FOR LOC ARE TAKEN FROM   ABRAHAMS WRITE-UP.
      LOC(1,1,1)=0
      LOC(1,1,3)=0
      LOC(1,1,2)=1
      LOC(1,1,6)=1
      LOC(1,1,7)=1
      LOC(1,1,4)=-1
      LOC(1,1,5)=-1
      LOC(1,1,8)=-1
      LOC(2,1,2)=0
      LOC(2,1,4)=0
      LOC(2,1,1)=1
      LOC(2,1,6)=1
      LOC(2,1,5)=1
      LOC(2,1,3)=-1
      LOC(2,1,7)=-1
      LOC(2,1,8)=-1
      DO 10 I=1,2
      DO 9 J=1,3
      DO 8 K=1,8
    8 LOC(I,J+1,K)=ROTATE(LOC,I,J,K)
    9 CONTINUE
   10 CONTINUE
      RETURN
      END
```

```
      FUNCTION  ROTATE (LOC,I,J,K)
      DIMENSION LOC(2,4,1)
      KSAVE=K
C     THIS FUNCTION PRODUCES THE VALUES OF LOC(I,J+1,K)  WHICH IS OBTAINED
C     BY ROTATING THE SITES THROUGH 90 DEGREES.
      IF(K-4) 1,2,3
    1 K=K+1
      GO TO 6
    2 K=1
      GO TO 6
    3 IF(K-8) 1,4,4
    4 K=5
      GO TO 6
    6 ROTATE=LOC(I,J,K)
      K=KSAVE
      RETURN
      END



      SUBROUTINE SETPAT(PATERN)
      INTEGER  PATERN(3,3,1)
C     THIS SUBROUTINE  STORES THE PROPER VALUES OF IROTATE IN THE VARIOUS
C     LOCATIONS OF THE PATTERN RECOGNITION CUBE, PATERN.
      PATERN(2,2,1)=1
      PATERN(1,2,2)=1
      PATERN(2,3,2)=2
      PATERN(3,2,2)=3
      PATERN(2,1,2)=4
      PATERN(1,1,3)=1
      PATERN(1,3,3)=2
      PATERN(3,3,3)=3
      PATERN(3,1,3)=4
      PATERN(2,1,4)=1
      PATERN(1,2,4)=2
      PATERN(2,3,4)=3
      PATERN(3,2,4)=4
      PATERN(2,2,5)=1
      RETURN
      END



      SUBROUTINE  MINORM(R,P,PU,SUM)
      DIMENSION  R(6,1),P(1)
C     THE PURPOSE OF THIS SUBROUTINE IS TO CALCULATE THE LARGEST SUM OF
C     RATES ENVOLVING UN-NORMALIZED P(I) AND R(I,J) AND PU THAT WILL EVER
C     ENCOUNTERED IN MAKING A DECISION CONCERNING CONDENSATION ,EVAPORATION
C     OR MIGRATION.  THE LARGEST SUM ,WHEN FOUND, WILL BE USED TO NORMALIZE
C     ALL  THE RATES P(I),R(I,J) AND PU.  THE EFFECTS OF SECOND NEAREST
C     NEIGHBORS ON TRANSITION RATES ARE NOT INCLUDED IN THIS SUBROUTINE
C     BUT THIS DOES NOT PREVENT THE MAIN PROGRAM FROM INCORPORATING THESE
C     EFFECTS AS LONG AS ALL  THE BONDING INVOLVED IS ATTRACTIVE.
C     CONSIDER THE CASE WHERE A PARTICLE HAS NO NEAREST NEIGHBORS.
C     THE RANDOM NUMBER GENERATOR  MUST SELECT BETWEEN EVAPORATION AND THE
C     EIGHT POSSIBLE LOCATIONS TO WHICH A PARTICLE MIGHT MIGRATE.
C     THE SUM OF THE RATES ASSOCIATED WITH THESE NINE PROCESSES IS LARGER
C     THAN THE SUM OF THE RATES THAT RESULT WHEN FIRST OR SECOND NEIGHBORS
C     EXIST ( AS LONG AS ALL ENERGIES OF INTERACTION ARE NEGATIVE).
C     WE ASSUME THAT ALL ENERGIES OF INTERACTION ARE NEGATIVE.
      SUM=P(1)
      DO 1 I=1,8
    1 SUM=SUM+R(1,I)
C     IF THIS SUM IS GREATER THAN PU ,THEN ALL RATES WILL  BE DIVIDED BY SUM.
C     IF PU IS LARGER, THEN ALL RATES WILL BE DIVIDED BY PU.
      IF(PU-SUM) 3,3,2
    2 SUM=PU
    3 DO 4 I=1,6
      DO 4 J=1,8
    4 R(I,J)=R(I,J)/SUM
      DO 5 I=1,5
    5 P(I)=P(I)/SUM
      PU=PU/SUM
      RETURN
      END



C     MAIN PROGRAM
C         .
C         .
```

```
C        .
C
C        THE FOLLOWING MAIN PROGRAM SEGMENT DOES NOT PERFORM ALL FUNCTIONS
C        DESCRIBED IN THE TEXT. IT DOES, HOWEVER, CONTAIN MOST OF THE
C        KEY FEATURES OF THE PROGRAM.
C
C        .
C        .
C        THE SITE IS OCCUPIED. FIND THE NUMBER OF NEAREST NEIGHBORS.
C        THE SECTION FROM HERE TO 138 COUNTS THE NUMBER OF NEAREST
C        NEIGHBORS OF SITE IX,IY AND STORES THE NUMBER IN N.
C        IXX AND IYY ARE INDICES ALONG WITH N  THAT GIVE LOCATIONS IN THE
C        PATTERN RECOGNITION CUBE.  SEE WRITE-UP.
C        IF SITE X,Y IS IN THE MIDDLE OF THE LATTICE , ITS NEAREST NEIGHBORS
C        HAVE INDICES THAT DIFFER FROM SITE X,Y ONLY BY 1 .
         IXM1=IX-1
         IXP1=IX+1
         IYM1=IY-1
         IYP1=IY+1
C        IF SITE X,Y HAPPENS TO LIE ON THE LATTICE BOUNDARY, SOME NEAREST
C        NEIGHBORS WILL HAVE INDICES THAT DIFFER BY MTXSIZ (MATRIX SIZE).
         IF(IXM1)100,100,101
   100   IXM1=MTXSIZ
   101   IF(IXP1-MTXSIZ)103,103,102
   102   IXP1=1
   103   IF(IYM1)104,104,105
   104   IYM1=MTXSIZ
   105   IF(IYP1-MTXSIZ)107,107,106
   106   IYP1=1
   107   N=0
C        THE FOLLOWING ST&TEMENTS 1 THRU 8          COUNT  THE NUMBER OF
C        NEAREST NEIGHBORS.
C        ADDRESSING IS ALSO SET UP FOR THE PATTERN RECOGNITION CUBE.
         IXX=2
         IYY=2
         IF(SITE(IXM1,IY))112,112,111
   111   N=N+1
         IXX=IXX-1
   112   IF(SITE(IXP1,IY))114,114,113
   113   N=N+1
         IXX=IXX+1
   114   IF(SITE(IX,IYM1))116,116,115
   115   N=N+1
         IYY=IYY-1
   116   IF(SITE(IX,IYP1))118,118,117
   117   N=N+1
         IYY=IYY+1
   118   CONTINUE
         IF(N-2)121,120,121
   120   IF(IXX-2)121,122,121
   121   NSTATE=N+1
         IROTAT=PATERN(IXX,IYY,NSTATE)
         GO TO 130
   122   NSTATE=6
         IF(SITE(IXM1,IY))123,123,124
   123   IROTAT=1
         GO TO 130
   124   IROTAT=2
   130   CONTINUE
C        THE NUMBER OF SECOND NEAREST NEIGHBORS ARE CALCULATED AND STORED IN N2.
         N2=0
         IF(SITE(IXP1,IYP1))132,132,131
   131   N2=N2+1
   132   IF(SITE(IXM1,IYM1))134,134,133
   133   N2=N2+1
   134   IF(SITE(IXP1,IYM1))136,136,135
   135   N2=N2+1
   136   IF(SITE(IXM1,IYP1))138,138,137
   137   N2=N2+1
   138   CONTINUE
C        THERE ARE N NEAREST NEIGHBORS AND N2 SECOND NEAREST NEIGHBORS.
         J=IRA*65539
         IF(J)1051,1052,1052
  1051   J=J+2147483647+1
  1052   RANDN=J
         RANDN=RANDN*.4656613E-9
         IRA=J
C        THE PROGRAM SEGMENT FROM HERE TO 229 IS CALLED BSCAN.
C        BSCAN  SCANS BINS DETWEEN 0 AND 1 TO FIND THE TRANSITION THAT THE
C        RANDOM NUMBER GENERATOR HAS SELECTED.
C        THE SIZE OF THE BINS IS DETERMINED SOLELY BY R(1)+R(2)+---R(6)+EVAP
C        AS SHOWN BELOW.
C        ONCE  THE TRANSITION HAS BEEN SELECTED, IT IS EXECUTED .
C        IROTAT AND NSTATE ARE GENERATED IN SUBROUTINE SEARCH.
C        N IS THE NUMBER OF NEAREST NEIGHBOR SITES.
```

```
C        N2 IS THE NUMBER OF SECOND NEAREST NEIGHBOR SITES.
C        NSTATE IS ONE OF THE SIX STATES DESCRIBED IN  NOTES.
C        R,P2,P ARE NORMALIZED  TRANSITION RATE VECTORS, WHICH ARE DESCRIBED
C        IN OTHER SUBROUTINES.
         EVAP=P(N+1)*P2(N2+1)*ENERGY(IX,IY)
         IF(RANDN-EVAP)210,201,201
     201 RBIN=EVAP
         DO 202 I=1,8
         RBIN=RBIN+R(NSTATE,I)*ENERGY(IX,IY)/ENERGY(1,2)
         IF(RANDN-RBIN)205,202,202
     202 CONTINUE
C        THE PARTICLE DOES NOT EVAPORATE OR MIGRATE . RETURN
         GO TO 229
     205 CONTINUE
C        THE PARTICLE MIGRATES TO LOCATION IX+LOC(1,IROTATE,I) AND  IY+
C        LOC(2,IROTAT,I).
    4000 GO TO(4001,4002,4003),NSNAPS
    4001 NSNAPS=3
         CALL DISPLA(MTXSIZ,IGDS2,NO,IATL,INTCD,IWENT)
         GO TO 4002
    4003 NSNAPS=1
         CALL DISPLA(MTXSIZ,IGDS1,NO,IATL,INTCD,IWENT)
    4002 CONTINUE
         IXSAVE=IX
         IYSAVE=IY
         SITE(IX,IY)=0
         IX=IX+LOC(1,IROTAT,I)
         IY=IY+LOC(2,IROTAT,I)
         IF(IX)220,220,221
     220 IX=MTXSIZ
         GO TO 223
     221 IF(MTXSIZ-IX)222,223,223
     222 IX=1
     223 IF(IY)224,224,225
     224 IY=MTXSIZ
         GO TO 227
     225 IF(MTXSIZ-IY)226,227,227
     226 IY=1
     227 CONTINUE
         IF(SITE(IX,IY))211,212,211
     211 SITE(IXSAVE,IYSAVE)=1
         GO TO 229
     212 SITE(IX,IY)=1
         GO TO 229
     210 CONTINUE
C        THE PARTICLE EVAPORATES.
         NUMPAR=NUMPAR-1
         SITE(IX,IY)=0
     229 CONTINUE
         GO TO 4
      10 CONTINUE
         NUM=NUM+1
         THETA=THET/(LOOPN*MTXSIZ*MTXSIZ)
     505 FORMAT(1H0'        THETA=',F10.5////)
         POINTS(NUM)=THETA
         IF(NUM-NUMCYC) 18,11,11
      18 ICOUNT=0
         THET=0
         GO TO 4
      11 TTIME=NUMCYC*LOOPN
         WRITE(6,506) TTIME
     506 FORMAT(1H0' THE TOTAL NUMBER OF SITES CHECKED IS ',F10.2)
C        CALCULATE  THETA-BAR BUT DO NOT INCLUDE THETA VALUES FROM THE
C        FIRST NN CYCLES BECAUSE EQUILIBRIUM IS NOT ATTAINED FOR THE
C        FIRST NN CYCLES.
C        .
C        .
C        .
         END
```

Chapter 5

Introduction to Computational Theoretical Chemistry

Robert N. Kortzeborn

IBM Scientific Center
Palo Alto, California

I. BASIC CONCEPTS OF COMPUTATIONAL THEORETICAL CHEMISTRY

At this time there is every reason to believe that all chemistry is deducible from the laws of quantum mechanics. But since quantum mechanics is inherently a mathematical discipline, the chemist who is interested in "real chemistry" tends to shy away from its study. Chemistry, with all its divisions and subdivisions, such as organic, inorganic, physical, and analytical, is further basically divided into experimental and theoretical branches of study.

In very general terms, both the theorist and experimentalist attempt to explain physical phenomena and predict events with "laboratory tools." The theoretical laboratory may consists of sharp pencils, reams of paper, and time on a large digital computer. Theoretical tools are used in conjunction with the more conventional laboratory apparatus of the experimental chemist—each providing complementary endeavors to ensure mutual impetus and augmentation. The experimentalist tends to work with "real problems" (i.e., chemical synthesis, analysis, etc.), while the theorist often is interested in a model of a system. These models are made mathematically tractable and represent reality only insofar as the model itself. The assumptions sometimes necessary to invoke in order to make a problem tractable often disguise results. Even after the assumptions are built into the model (allowing the solution to be attempted), one is left with many problems ranging from large amounts of computer time necessary to carry the problem to "solution" to completeness of the model. However, even within these limitations astounding progress is being made at present.

In the first 10 years of its application, quantum mechanics was used to solve many evasive problems. The periodic system of the elements, the covalent bond, the aromatic character of benzene, the existence of free radicals, van der Waals forces, the magnetic properties of matter, and the conduction of electricity by metals are a few of the phenomena finally explained in terms of quantum mechanics. This has led people to believe that a new age of chemistry is at hand, one in which a chemist can resolve unanswered questions with a computer. However, *accurate* (in the sense of complete) quantum-mechanical calculations on all but the most simple systems are well beyond the best computers. Also, many problems that were expected to be vulnerable to quantum-mechanical interpretation await other methods for their solution (e.g., the structure of boron hydrides and properties of optical isomers).

Very few chemical concepts can adequately be understood without grasping the basics of quantum mechanics (this includes the chemical bond). The chemist should be aware of the limitations of the field and be sure that he is not deluded into thinking that the concepts and results are better (or more valid) than they really are. In learning to make the necessary mathematical manipulation of quantum mechanics, the chemist should not be misled into believing that he is necessarily acquiring a tool for solving experimental problems.

However, within these contraints, we are interested in answering the question posed by Noble laureate chemist Robert S. Mulliken: "What are the electrons really doing in molecules?"[1] This question defines the heart of chemical research and describes the quantum theory of valence.[2-5] High-speed digital computers have had a profound affect on chemistry and physics and in particular on the quantum theory of valence. We are now able to attempt to accurately describe certain chemical bonds.[6]

The problems associated with describing a large molecule or large systems of large molecules also are being met.[7] Theories that break down large molecules into smaller tractable components are currently under investigation. The theoretical approach to small molecular systems is having immediate and direct effect on the theory of large organic molecules. At one time it was feared that large molecules were inaccessible to accurate computational treatment because they contained too many electrons. This fear is currently being dispelled through concentrated research on small-

and medium-sized molecules. The research does not just report the results of a quantum-mechanical description (viz., the molecular wave function) but rather uses these wave functions to compute physically observable quantities that can be checked via experiment.[8-10] That is, computational theoretical chemistry is using the results of long arduous calculations (the wave functions) in order to report physical observables rather than just reporting accurate wave functions that require many pages of quite incomprehensible computer output.

The field of computational science is growing at an accelerating rate. Little of an extremely diverse field will even be mentioned here. Investigations into large molecules of biological significance now comprise an active research endeavor. The tools and results of molecular quantum mechanics will be increasingly useful to practicing organic, inorganic, and physical chemists as well as spectroscopists and biologists as the knowledge about molecules and their behavior grows. We have seen purely theoretical calculations used as standards for determining properties of small molecules. As quantum theory develops, theoretical calculations will become an acceptable[11] method for determining properties of small molecules. The time is approaching when the chemist will be able to access large precomputed files of information via time-shared computer terminals in order to perform an experiment rather than having to go into his laboratory. If the restrictions of the mathematical models are kept in mind while using computer models, quantum theory will find far-reaching and bountiful applications in chemistry.

In summary then, insofar as quantum mechanics is correct, chemical questions are problems in applied mathematics. Because of its complexity, chemistry will not cease, for a large part, to be a marriage of theory and experiment. No chemist can afford to be uninformed about a theory that systematizes all of chemistry even though the mathematical complexities are, at times, trying. However, even to the "elite, sophisticated" mathematical chemical physicist, one must allude to a quote by Albert Einstein: "Insofar as mathematics applies to reality, it is not certain, and so far as mathematics is certain it does not apply to reality."

II. THE NATURE OF THE PROBLEM

During the late 1920's, Erwin Schrödinger wrote down a time-independent differential equation resulting from the just discovered discipline called "quantum mechanics." This equation is simply written as

$$H\Psi = E\Psi \tag{2.1}$$

The only difficulty in relating this equation to physical reality is in the mathematics. It is extremely difficult to solve for systems of physical interest. In Eq. (2.1), H is the Hamiltonian operator for the system, Ψ is a "wave function" and E is the energy for the system described by Ψ. The wave function, Ψ, has no physical meaning; however, its absolute square, $|\Psi|^2$, measures the probability of finding the system in a given configuration. Equation (2.1) is an eigenequation. That is, when Eq. (2.1) is satisfied, E is an energy eigenvalue and Ψ is an energy eigenfunction. (For a review of eigen problems, see Kauzmann.[12]) This equation when coupled with the Pauli exclusion principle has quantitatively explained energy levels of light elements where relativistic effects are unimportant.*

The solution of Eq. (2.1) is usually attempted by some method of successive approximations. If we approximate Ψ by φ, the solution of Eq. (2.1) is simply

$$E = \int \varphi^* H\varphi \, d\tau \Big/ \int \varphi^* \varphi \, d\tau \tag{2.2}$$

where the integral is overall the configurational space of the system, $d\tau$ is the volume element including particle spin, and the asterisk denotes the complex conjugate of the function.

We now employ a procedure called the "variational method" for obtaining the approximate ground state wave function and energy. The development and properties of this theory can be found in many elementary texts on quantum mechanics.[13] Due to space limitations, we shall only consider its use, not its development. Once we know the electronic wave function (the eigenfunction) as obtained from the variational procedure, we can compute other properties of the system as well because any physical observable average value whose operator is known may be computed in theory by the following equation:

$$x = \int \Psi^* (OP_x) \Psi \, d\tau \Big/ \int \Psi^* \Psi \, d\tau \tag{2.3}$$

where x is the average value of the variable whose operator is (OP_x). It may well happen that if the wave function does not represent the real function adequately (i.e., span the configuration space adequately), the energy may be computed but the function will not predict other properties very

---

* The relativistic effects are not important for light elements. This effect is more important for elements of large atomic number. See Noble and Kortzeborn[9] and the references cited therein for estimation procedure details.

well. Of course, as the wave function is refined in order to span the configuration space, it will ultimately become adequate for a wider range of properties.

Usually the Hamiltonian for a given system can be written down by inspection (if we ignore small interaction terms). The real problem is then:

1. Guessing at suitable trial electronic wave functions φ (and hence Ψ).
2. Calculating the energy and other average values of interest.
3. Changing the parameters in φ such that E gets smaller and approaches a true value and iterate until we obtain a satisfactory convergence. (E gets smaller and approaches a true limit by virtue of the variational principle. See Kauzmann[12] for details.)

In order to obtain a better feel and to grasp the nature of the problem, we shall consider some very simple atomic cases. These consist of the hydrogen atom (whose solution is exact) and the helium atom (whose solution is approximate in the mathematical sense).

A. The Hydrogen Atom (Ground Electronic State)

We shall now treat the simplest atomic system (viz., the hydrogen atom) by the methods of quantum mechanics. This atom consists of two fundamental particles, an electron of charge $-e$ and mass m and a proton of charge $+e$ and mass M. The particles attract each other according to the Coulomb law of electrostatic interaction. The potential energy of this system (V) is simply

$$V = \frac{-e^2}{[(x_1 - x_2)^2 + (y_1 - y_2)^2 + (z_1 - z_2)^2]^{\frac{1}{2}}} \qquad (2.4)$$

where x_1, y_1, and z_1 denote the coordinates of the proton and x_2, y_2, and z_2 denote the coordinates of the electron. This form of the potential energy leads to a complicated form for the Hamiltonian operator [viz., H in Eq. (2.1)]. The Hamiltonian operator is the sum of the kinetic energy and potential energy and can be written

$$H = -\frac{1}{2}\nabla^2 + V \qquad (2.5)$$

where ∇^2 is the well-known Laplacian operator. If we transform coordinates to a center of mass system we obtain, after some manipulation, separable equations that can be solved in closed form. Using Eqs. (2.4) and (2.5) and solving Eq. (2.2) for the energy eigenvalue E, one obtains

$$E = -\frac{2\pi^2 \mu e^4}{n^2 h^2} \qquad n = 1, 2, 3, \ldots \qquad (2.6)$$

where $\mu = mM/(M + m)$, which is the so-called "reduced mass" of the system, h is Planck's constant, and n is the principal quantum number.

The assumptions necessary to solve Eq. (2.2) for this system are not presented here in order not to burden the reader with unnecessary detail. These details can be found in numerous elementary quantum-mechanics texts.[13,14]

For "real life" atomic and molecular systems, the situation is not at all this simple. The solution to Eq. (2.1) is not possible in closed form, so more approximate techniques are employed. As a matter of fact, the only real atomic system that can be solved *exactly* in closed form is the hydrogen atom. Any system with more than one electron can only be solved by approximate techniques. We shall illustrate this by considering the simplest atomic system that contains more than one electron, i.e., the helium atom. (See Section IIB.)

Before pressing on to that solution, however, a few comments concerning the Hamiltonian operator are in order. In almost all cases, one knows how to write down the Hamiltonian operator based on the classical energy expression. The resultant integral equations are very hard and complicated. The real Hamiltonian operator for a system consists of *many* terms:

$$H = \sum_{\substack{\text{electrons} \\ (j)}} \frac{P_j^2}{2m_j} + \sum_{\substack{\text{nuclei} \\ (K)}} \frac{P_K^2}{2m_K} + \sum_{\substack{\text{nuclei-} \\ \text{electrons} \\ (K,j)}} \frac{Z_K e^2}{r_{Kj}} + \sum_{\substack{\text{nuclei} \\ (K,L)}} \frac{Z_K Z_L}{r_{KL}} + \cdots$$

$$+ \sum_{\substack{\text{electrons} \\ (i \neq j)}} \frac{e^2}{r_{ij}} + H_{so} + H_{ss} + H_{hfs} + H_{\text{ext fields}} + \cdots \qquad (2.7)$$

In Eq. (2.7) the first term represents the kinetic energy of the electrons. P_j represents the momentum of the electrons and, in practice, is replaced by the Laplacian operator. In Cartesian coordinates it is simply

$$\nabla^2 = \frac{\partial^2}{\partial x^2} + \frac{\partial^2}{\partial y^2} + \frac{\partial^2}{\partial z^2} \qquad (2.8)$$

The second term in Eq. (2.7) represents the kinetic energy of the nuclei with momentum P_K. The Laplacian operator is again used in practice (see Kauzmann[12] for details of energy operators and the corollaries of quantum mechanics). The third, fourth, and fifth terms represent potential energy terms between the nuclei and electrons, the nuclei with other nuclei, and the electron–electron terms, respectively. The remaining terms of H_{so}, H_{ss}, and H_{hfs} refer to the spin–orbit, spin–spin, and hyperfine structure couplings. The solution to Eq. (2.1) is the solution to an "eigenvalue" problem where the eigenfunctions will be functions of all the variables that enter into the problem. Therefore the eigenfunctions are more than $3N$-dimensional, where N is the number of electrons. Complete solutions cannot be expected then, except for "textbook cases" such as harmonic oscillators, single electron central field problems (the hydrogen atom), and square-walled boxes.[15]

The real solution to real problems cannot even be accomplished by using high-speed computers. If we wanted only to write down the resulting wave function (a $3N$-dimensional function), we must divide each axis into, say, 100 units (for some accuracy). So we only need $100^{3N} = 10^{6N}$ entries to tabulate the wave function. If we consider a simple four-particle system we have 10^{24} entries! This is more than Avagadro's number and would require millions of volumes to record the result. The human mind would have "some trouble" sorting through these numbers in order to find some meaning.

The only practical method of approach is therefore approximate in nature. One approaches the problem by including more and more small terms in the Hamiltonian and refining the results. This procedure can yield satisfactory results in certain cases. In order to put things in a more quantitative light, let us consider the following case.

B. The Helium Atom (Ground Electronic State)

We shall now make use of both simple perturbation and variation theory. We can use these techniques to approximate solutions for the helium atom in its ground electronic state. Perturbation and variation theory play important roles in quantum mechanics. Since this is meant to be an introduction to computational chemistry, we will use very simple results with no derivations in order to expedite our main theme. The reader is referred to References 13 to 18 for the details of perturbation theory. We shall use only the simple "first-order" results in this section. An excellent treatise concerning perturbation theory and its application to quantum-mechanical problems may be found in Reference 18.

The Hamiltonian operator for the helium atom (or for other two-electron atoms, such as Li$^+$) is given by Eq. (2.9)

$$H = -\frac{h^2}{8\pi^2 m}(\nabla_1^2 + \nabla_2^2) - \frac{Ze^2}{r_1} - \frac{Ze^2}{r_2} + \frac{e^2}{r_{12}} \qquad (2.9)$$

if we neglect the terms arising from the motion of the nucleus. Equation (2.9) includes the kinetic energy Laplacian operators for electrons one and two (i.e., ∇_1^2 and ∇_2^2), the potential energy terms of electrons one and two with the nucleus of charge Z (i.e., $-Ze^2/r_1$ and $-Ze^2/r_2$), and finally the e^2/r_{12} that represents the electron–electron potential. This electron–electron potential gives rise to many problems in theoretical chemistry. We shall address the details of some of these problems later. Perturbation theory will now be used to examine the helium atom Hamiltonian.

For the purpose of simplifying the calculation of the integrals involved in problems of this type it is usually more convenient to use atomic units. The transformation to atomic units is obtained by expressing the distances in terms of Bohr radius. We define the Bohr radius a_0 as

$$a_0 = \frac{h^2}{4\pi^2 me^2} \qquad (2.10)$$

The final Hamiltonian operator that results in atomic units is given by Eq. (2.11) (in units of e^2/a_0):

$$H = -\frac{1}{2}(\nabla_1^2 + \nabla_2^2) - \frac{Z}{R_1} - \frac{Z}{R_2} + \frac{1}{R_{12}} \qquad (2.11)$$

where we use R_i instead of r_i to denote atomic units.

One now applies the simple methods of perturbation theory to the problem. We divide the Hamiltonian operator into two operators, viz.,

$$H = H_0 + H^{(1)} \qquad (2.12)$$

where

$$H_0 = -\frac{1}{2}(\nabla_1^2 + \nabla_2^2) - \frac{Z}{R_1} - \frac{Z}{R_2} \qquad (2.13)$$

and

$$H^{(1)} = \frac{1}{R_{12}} \qquad (2.14)$$

($H^{(1)}$ is the first-order perturbation Hamiltonian.) The zeroth-order eigenfunctions are solutions of the equation

$$H_0 \Psi^0 = E^0 \Psi^0 \tag{2.15}$$

Now we set $\Psi^0 = \Psi^0(1)\Psi^0(2)$ and $E^0 = E^0(1) + E^0(2)$ [(1) and (2) denote electrons (1) and (2), respectively]; Eq. (2.15) is separable into two equations:

$$\frac{1}{2} \nabla_1^2 \Psi^0(1) + \left(E^0(1) + \frac{Z}{R_1} \right) \Psi^0(1) = 0$$

$$\frac{1}{2} \nabla_2^2 \Psi^0(2) + \left(E^0(2) + \frac{Z}{R_2} \right) \Psi^0(2) = 0 \tag{2.16}$$

Note that these two equations are just those for a hydrogen-like atom with a nuclear charge of Z (see Section IIA). For the ground state of the helium atom, one therefore uses the normalized functions that are known from an analysis of the hydrogen atom. These are[13]

$$\Psi^0(1) = \frac{1}{\sqrt{\pi}} Z^{3/2} e^{-ZR_1} \qquad \Psi^0(2) = \frac{1}{\sqrt{\pi}} Z^{3/2} e^{-ZR_2}$$

with

$$\Psi^0 = \Psi^0(1)\Psi^0(2) = \frac{Z^3}{\pi} e^{-Z(R_1+R_2)} \tag{2.17}$$

and

$$E^0 = E^0(1) + E^0(2) = 2Z^2 E(H)$$

where $E(H)$ is the ground state energy (i.e., lowest energy) of the hydrogen atom and is in units of e^2/a_0 given by $-e^2/2a_0$.

Now the general first-order perturbation energy correction is simply

$$E^{(1)} = \iint \Psi^{0*} H^{(1)} \Psi^0 \, d\tau_1 \, d\tau_2$$

$$= \frac{e^2}{a_0} \frac{Z^6}{\pi^2} \int_{R_1} \int_{R_2} \frac{e^{-2ZR_1} e^{-2ZR_2}}{R_{12}} \, d\tau_1 \, d\tau_2 \tag{2.18}$$

(Ψ^{0*} is the complex conjugate of Ψ^0), where the volume elements of integration for electrons one and two are denoted by $d\tau_1$ and $d\tau_2$, respectively. In terms of spherical polar coordinates (R, θ, and φ coordinates) the volume elements are

$$d\tau_1 = R_1^2 \sin \theta_1 \, dR_1 \, d\theta_1 \, d\varphi_1$$

$$d\tau_2 = R_2^2 \sin \theta_2 \, dR_2 \, d\theta_2 \, d\varphi_2 \tag{2.19}$$

We are now faced with a small dilema. The indicated integration in Eq. (2.18) cannot be performed because we must integrate overall space. That is, R_{12} is just the distance between electron one and electron two, viz.,

$$R_{12} = \mid \mathbf{R}_1 - \mathbf{R}_2 \mid$$

It is obvious that R_{12} can be zero and therefore the integral in Eq. (2.18) can be infinite (i.e., the integral contains a pole). We can circumvent this snag by invoking a well-known expansion theorem.[13] The quantity $1/R_{12}$ can be expanded in terms of associated Legendre polynomials as

$$\frac{1}{R_{12}} = \sum_{l}^{\infty} \sum_{m=-l}^{m=+l} \frac{(l - |m|)!}{(l + |m|)!} \frac{R_<^l}{R_>^{l+1}} P_l^{|m|}(\cos\theta_1) P_l^{|m|}(\cos\theta_2)$$

$$\times \exp\{im(\varphi_1 - \varphi_2)\} \tag{2.20}$$

In Eq. (2.20), $R_<$ and $R_>$ are the lesser and greater of the quantities R_1 and R_2. $P_l^{|m|}(\cos\theta)$ are the associated Legendre polynomials whose properties can be found in many elementary references.[13] The quantities l and m are integer expansion parameters and correspond to two distinct quantum numbers in the canonical representation of hydrogenic wavefunctions. The subscripts 1 and 2 correspond to electron 1 and 2, respectively, and the angles θ and φ correspond to the well-known spherical polar coordinate system.

Note that the wave functions we are using [cf. Eq. (2.17)] do not contain functions of angles. We obtained the functions in Eq. (2.17) from inspection of Eq. (2.16). We could have chosen more complete functions to represent the solution by including, as a multiplicative factor, the spherical harmonic angular functions. This point will be addressed in Section III in some detail. In general, wave functions indeed do contain angular functions that give rise to many computational problems.

In our present computation, however, we need not be concerned with these problems. The associated Legendre functions happily display some rather convenient properties, one of which is orthogonality. This simply means that when we use Eq. (2.20) in Eq. (2.18) all summations vanish except for those where $l = 0$ and $m = 0$. For these terms, $P_0(\cos\theta) = 1$, so Eq. (2.18) simply reduces to

$$E^{(1)} = \frac{e^2}{a_0} \frac{Z^6}{\pi^2} \int_{R_1} \int_{R_2} \frac{e^{-2ZR_1} e^{-2ZR_2}}{R_>} d\tau_1 \, d\tau_2 \tag{2.21}$$

The angular integration from $d\tau_1$ and $d\tau_2$ gives a factor of $4\pi^2$, so we have only the integral over R_1 and R_2, which may be written

$$E^{(1)} = 16Z^6 \frac{e^2}{a_0} \int_0^\infty e^{-2ZR_1} \left[\frac{1}{R_1} \int_0^{R_1} e^{-2ZR_2} R_2^2 \, dR_2 \right.$$

$$\left. + \int_{R_1}^\infty e^{-2ZR_2} R_2 \, dR_2 \right] R_1^2 \, dR_1 \tag{2.22}$$

which is easily evaluated to give

$$E^{(1)} = \frac{5}{8} Z \frac{e^2}{a_0} \tag{2.23}$$

To the first-order perturbation approximation, then, the energy of He or helium-like atoms is

$$E = E^0 + E^{(1)} = \left(2Z^2 - \frac{5}{4} Z \right) \left(-\frac{1}{2} \frac{e^2}{a_0} \right)$$

$$= \left(2Z^2 - \frac{5}{4} Z \right) E(H) \tag{2.24}$$

The energy of the ground state of He$^+$ is $Z^2 E(H)$. The energy necessary to remove one electron is

$$\left(Z^2 - \frac{5}{4} Z \right) E(H) = \frac{3}{2} E_{1s}(H) = \frac{3}{2} (13.60) = 20.4 \text{ eV}$$

The observed value is 24.58 eV, so our calculated value is off by $\sim 16\%$. With more detailed calculations, we can reduce this error to essentially zero.

The mathematics for more complex atomic systems is essentially the same as put forth here. When we examine the more complex molecular systems, integrals arise that involve the R_{12}^{-1} terms explicitly. The integrals become much more complicated because they are functions of up to four atomic centers and two electrons. The resultant solutions are both more difficult and time-consuming to obtain.

III. REAL MOLECULAR SYSTEMS

A. Introduction

The quantum-mechanical analysis of real molecular systems is a problem that makes use of many approximate techniques by necessity. In the next section we take a short look at one successful approximate technique called the "Hartree–Fock model." Here we shall examine the type of approximate analytical functions used and note the time-consuming step in the computation.

Two types of analytical functions in current use are Gaussian-type functions (GTO) and Slater-type functions (STO). The Gaussian-type functions are defined as

$$
\begin{aligned}
\chi(A, \alpha, l, m, n) &= (A, \alpha, l, m, n) \\
&= (X - A_x)^l (Y - A_y)^m (Z - A_z)^n \exp(-\alpha r_A^2) \\
&= X_A^l Y_A^m Z_A^n \exp(-\alpha r_A^2)
\end{aligned}
\tag{3.1}
$$

Here A denotes the atomic center A in the molecule; X, Y, Z denote electronic coordinates; A_x, A_y, A_z denote the coordinates of center A relative to some origin; and α is an "orbital exponent."

We want to normalize the wave function to unity. That is,

$$
\int \Psi^* \Psi \, d\tau = 1
\tag{3.2}
$$

The normalizing factor that accomplishes this for these GTO is

$$
N\alpha = \left[\left(\frac{\pi}{2\alpha} \right)^{\frac{3}{2}} \frac{(2l-1)!!(2m-1)!!(2n-1)!!}{2^{2(l+m+n)} \alpha^{(l+m+n)}} \right]^{\frac{1}{2}}
\tag{3.3}
$$

Rather than go into all the details, we simply write down two typical integrals that must be evaluated. Let us denote the integrals by K and J:

$$
K_{ijkl}^{ABCD} = \int\int \frac{\chi_i^A(1)\chi_j^B(1)\chi_k^C(1)\chi_l^D(2)}{r_{12}} \, d\tau_1 \, d\tau_2
\tag{3.4}
$$

$$
J_{ijkl}^{ABCD} = \int\int \frac{\chi_i^A(1)\chi_j^B(1)\chi_k^C(2)\chi_l^D(2)}{r_{12}} \, d\tau_1 \, d\tau_2
\tag{3.5}
$$

where K is an "exchange integral" and J is a so-called "Coulomb integral." They involve GTO's on four different centers and the r_{12}^{-1} operator. The superscripts A, B, C, and D refer to the atomic centers; the subscripts i, j, k, and l refer to function i, j, k, or l centered on atoms A, B, C, or D, since we can (and do) have many functions on the same center. Finally, the numbers in parentheses refer to electrons. This is only done for convenience since we cannot really distinguish between electrons. This indistinguishability leads to other types of integrals, which we shall leave out of the discussion. The volume elements $d\tau_1$ and $d\tau_2$ represent the volume elements associated with each electron. Integration may be performed over x_1, y_1, z_1, x_2, y_2, and z_2 (Cartesian coordinates) or r_1, θ_1, φ_1, r_2, θ_2, φ_2 (spherical polar coordinates), which leads in either case to a six-dimensional integration. Equations (3.4) and (3.5) do indeed possess a "pole" in the same sense we saw in the He atom discussion. Therefore, approximate techniques of evaluating these integrals are employed.

In a typical molecular computation, the calculation of integrals as in Eqs. (3.4) and (3.5) is the rate-determining step. If we use a total of 100 functions (i.e., $i + j + k + l = N = 100$), one needs to compute *roughly*

$$\frac{N^4}{8} = \frac{(100)^4}{8} \quad \text{integrals}$$

Since these integrals contain poles, one must use approximate time-consuming techniques in their evaluation. The use of GTO's in molecular computations grew out of a need for fast integral evaluation.

The second type of mathematical function "in vogue" today is the Slater-type orbital (STO). It is defined as (normalized)

$$\varphi_A{}^i(n, l, m, \alpha) = [(2n_i)]^{-\frac{1}{2}}(2\zeta)^{n_i+\frac{1}{2}}r^{n_i-1}e^{-\zeta_i r_i}Y_{lm}(\theta_i\varphi_i) \qquad (3.6)$$

Here r_i is the electronic coordinate on some center; n_i, l_i, and m_i are integer quantum numbers; ζ_i is an orbital exponent: and $Y_{lm}(\theta, \varphi)$ are spherical harmonics of either real or imaginary argument. The STO is a better function to describe real systems, but the integrals that arise [analogous to Eqs. (3.4) and (3.5)] for STO's are much more difficult to evaluate than the GTO expressions. Both types of functions are in current use. One model heavily exploited in computational chemistry is that due to Hartree and Fock. We now take a brief look at this model.

B. The Method of Hartree and Hartree-Fock

The wave function that would best describe a multielectron atom or molecule should include the distance between the electrons explicitly (viz., r_{ij}). But because of the great difficulties that would result from the use of such functions in many-electron atoms and molecules, we limit ourselves, by judicious choice, to products of one-electron functions that do not contain r_{ij} explicitly. The best possible such function would be obtained by application of the variation method as previously discussed. In the early work of Hartree, the simple product form of the one-electron function was used. We now examine this method in order to demonstrate one possible mathematical model. We desire to minimize the energy; in other words, we want to minimize the equation

$$E = \frac{\displaystyle\int \Psi^* H \Psi \, d\tau}{\displaystyle\int \Psi^* \Psi \, d\tau} \tag{3.7}$$

i.e., satisfy the equation

$$\delta E = \delta \int \Psi^* H \Psi \, d\tau = 0 \tag{3.8}$$

The above method of expressing the energy is valid only if the wave functions are normalized [cf. Eq. (3.2)]. We can ensure that this condition is fulfilled if we require that each atomic function is normalized. The Hamiltonian operator for the system of n electrons is

$$\begin{aligned} H &= \sum_{i=1}^{n} \left\{ -\frac{h^2}{8\pi^2 m} \nabla_i{}^2 - \frac{Ze^2}{r_i} \right\} + \frac{1}{2} \sum_i \sum_{j \neq i} \frac{e^2}{r_{ij}} \\ &= \sum_{i=1}^{n} H_i + \frac{1}{2} \sum_i \sum_{j \neq i} \frac{e^2}{r_{ij}} \end{aligned} \tag{3.9}$$

The solution of Eq. (3.8) is subject to the normalization condition and shows that the best possible wave function of the type of a product of atomic functions is obtained by using the set of atomic functions that are solutions to the n simultaneous equations

$$H_i \varphi_i + \left\{ \sum_{j \neq i} e^2 \int \frac{|\varphi_j|^2}{r_{ij}} \, d\tau_j \right\} \varphi_i = \varepsilon \varphi_i \tag{3.10}$$

Remember that H is a product of one-electron orbitals φ_k as

$$H = \varphi_1(1)\varphi_2(2) \cdots \varphi_n(n) \tag{3.11}$$

Now $e^2 |\varphi_j|^2$ is just the charge distribution of the jth electron and $e^2 |\varphi_j|^2 / r_{ij}$ is the potential energy of the ith electron in the field of the jth electron. The method of solution of the set of simultaneous equations is the following:

An arbitrary set of φ_i's is chosen with the choice being guided by any previous knowledge of approximate wave functions for the atom. The field arising from all the electrons, except the ith electron, is calculated from this set of φ's, and the φ_i is then calculated from Eq. (3.10). This procedure is carried out for each of the n electrons. Since the calculated set of φ's will not be identical with the original set, the calculations are repeated with a new set of φ's that have been augmented in a manner suggested by the results of the first calculation. This process is continued until the trial set of φ's and the calculated set are very nearly identical (to a preset tolerance), at which point the solution of the set of simultaneous equations has been achieved. The calculated charge distribution gives a set of φ's that will reproduce the real charge distribution quite well. Because of this iterative technique, one can understand why this method is frequently called the "method of self-consistent field."

The energy associated with the wave function of the type in Eq. (3.11) is a product of atomic functions and is given by

$$E = \int \Psi^* H \Psi \, d\tau$$

$$= \sum_{i=1}^{n} \varphi_i^* H \varphi_i \, d\tau_i + \frac{1}{2} \sum_i \sum_{j \neq i} e^2 \int\int \frac{|\varphi_i|^2 |\varphi_j|^2}{r_{ij}} \, d\tau_1 \, d\tau_2 \quad (3.12)$$

If the φ's are the solutions of Eq. (3.10), the energy will be given by

$$E = \sum_{i=1}^{n} \varepsilon_i - \frac{1}{2} \sum_i \sum_{j \neq i} e^2 \int\int \frac{|\varphi_i|^2 |\varphi_j|^2}{r_{ij}} \, d\tau_i \, d\tau_j \quad (3.13)$$

The reason there are no energy terms in Eq. (3.13) corresponding to the "exchange energy" [cf. Eq. (3.4)] is that the original wave function Ψ was written as a simple product rather than a determinant, as it should be if the Pauli exclusion principle is to be satisfied.[13] By writing the wave function as a determinant and proceeding as above with the additional requirements that the φ's be orthogonal, one is led to Fock equations.[14,19] These equations are similar in form to the Hartree equation (Eq. 3.10) but contain additional terms corresponding to the potential energies arising from the electron interchange.

Although the above wave functions, particularly those obtained by the solution of the Fock's equation, are the best possible one-electron wave functions, they give values for the energy of atoms that are incorrect by *approximately* 0.5 volt per electron.

When we couple the Hartree method with the method of Fock we obtain the Hartree–Fock model. The Hartree–Fock model contains both coulombic and "exchange terms" [cf. Eqs. (3.4) and (3.5)]. The Hartree–Fock model is being used today as a powerful research tool. It enables the theoretical chemist—armed with appropriate money, programs, and computers—to examine the molecules of interest. The analysis is made and the results checked against experiment in order to find wave functions that yield many physical observables simultaneously and thus yield information on the basic chemical bond.

Many other computational techniques and models are currently being used. The outline given here is only meant to acquaint the reader with one prevalent method. This method is being used today as only one basic research tool of theoretical chemistry.

C. Multicenter Integrals

The preceding pages of this manuscript are meant to provide a basic brief overview of computational chemistry. We must answer the following questions: How does one really attack a particular problem? How does one implement a theory using the computer? In order to illustrate one particular approach to the problem, the remainder of this chapter is devoted to a possible new method of computing multicentered, two-electron integrals that arise in molecular quantum mechanics. This method is presently in the research stage. Section IV explains the theoretical approach developed by R. N. Kortzeborn and W. D. Gwinn (Department of Chemistry, University of California, Berkeley, California). The procedure evaluates basic two-center, one-electron integrals over Slater-type functions and then uses these integrals to obtain the desired two-electron, multicentered integrals via transformation techniques.

We now present the analytical expressions used to evaluate an initial so-called "one-electron, two-centered integral" using the Slater-type orbitals (STO) of Eq. (3.6). As before, the integration is over all configurational space of the electron. If we let X_a represent the STO centered on center a, X_b represent the STO centered on center b, and M represent a general operator, we can write

$$(\chi_a \mid M \mid \chi_b) = \int_0^\infty \int_0^\pi \int_0^{2\pi} \chi_a M \chi_b r^2 \sin \theta \, dr \, d\theta \, d\varphi \qquad (3.14)$$

if we assume the usual spherical polar coordinates. The integral in Eq. (3.14) is now recast in the following equations, to a form more suitable for computational evaluation. The result is presented in Eqs. (3.15) to (3.19). Details may be found in Wahl et al.[20] We include these equations in order to give the reader an idea of the nature of a very simple such integral and in order to explicitly indicate the nature of the associated coding problems. Due to space limitations, we only include a partial FORTRAN code listing for the problem.* This code follows the explanation given in Section IV in detail. One is forced to do a good deal of "bookkeeping" in such codes because of the nature of the direct product transformations and the limited core size on most computers. These problems are somewhat alleviated now because of the advent of virtual computers. These machines allow one to code a problem that is essentially core contained and not worry about the bookkeeping associated with I/O problems. An explanation of the very powerful virtual computer concepts and their use in the scientific environment may be found in Kortzeborn.[25]

In computational terms, Eq. (3.14) is written

$$(\chi_a \mid M \mid \chi_b) = \delta(m_a, m_b + \sigma)[V_{2n_a}(\zeta_a R)V_{2n_b}(\zeta_b R)]^{-\frac{1}{2}}(\tfrac{1}{2}R)^{\lambda}$$

$$\times \sum_{t=0}^{t_{max}} Q_t \sum_{r=0}^{[\frac{1}{2}(l_a-m_a)]} \sum_{s=0}^{[\frac{1}{2}(l_b-m_b)]} W(r)W(s)$$

$$\times L_{n_a-l_a+2r+a_t,\ n_b-l_b+2s+b_t}^{l_a-m_a-2r+c_t,\ l_b-m_b-2s+d_t,\ \frac{1}{2}(m_a+m_b+e_t)}(\varrho, \tau) \qquad (3.15)$$

where

$$(\chi_a \mid M \mid \chi_b) = \int \chi_a M \chi_b \, d\tau \qquad (3.16)$$

and

$$V_N(X) = X^{-N-1}N! \qquad (3.17)$$

and

$$L_{\alpha\beta}^{\gamma\delta\varepsilon}(\varrho\tau) = \int_1^\infty d\xi \int_{-1}^1 d\eta (\xi + \eta)^\alpha (\xi - \eta)^\beta (1 + \xi\eta)^\gamma (1 - \xi\eta)^\delta$$

$$\times [(\xi^2 - 1)(1 - \eta^2)]^\varepsilon \exp(-\varrho\xi - \tau\varrho\eta) \qquad (3.18)$$

and

$$W(r) = \left[\frac{(2l_a + 1)}{2} \frac{(l_a - m_a)!}{(l_a + m_a)!}\right]^{\frac{1}{2}} \frac{(-1)^{m_a+r}(2l_a - 2r)!}{2^{l_a}(l_a - m_a - 2r)!(l_a - r)!l_a} \qquad (3.19)$$

where n_a, l_a, and m_a are quantum numbers for functions centered on a.

The inclusion of Eqs. (3.15) to (3.19) is to give the reader some "feeling" for a typical simple integral that indeed arises in the quantum mechanical

---

* Due to space limitations, only two FORTRAN routines have been included in the appendix. The first routine (MAIN) indicates the calling sequence of the other routines necessary to perform the complete computation. Subroutine TWO is included in order to indicate the code necessary to compute a two-center one-electron integral used as an "initial" function in the problem.

solution of the Schrödinger equation for a diatomic system. The theory then uses these integrals in a manner outlined in the following section to evaluate the much more complicated molecular integrals of Eqs. (3.4) and (3.5). These integrals then are used within the Hartree–Fock model [cf. Eqs. (3.7) to (3.13)] to obtain the solution to the problem (viz., Ψ and the associated energy).

IV. THE CALCULATION OF QUANTUM-MECHANICAL TWO-ELECTRON MULTICENTER INTEGRALS VIA TRANSFORMATION THEORY

A. Introduction

The use of transformation theory to calculate various quantum-mechanical matrix elements has proved to be an extremely powerful method when applied to one-dimensional problems. The complexity is increased in going to multidimensional problems, but it is still far simpler and faster to calculate the desired multidimensional expectation values via this method than via the conventional methods. In this section transformation theory will be used to calculate most of the three- and four-center electron repulsion integrals of molecular quantum mechanics.

B. General Theory

In molecular quantum-mechanical calculations for electronic energy levels, wave functions are usually expanded in a limited set of basis functions centered on each atom. These functions are usually the first few members of an infinite and complete set of functions spanning the space of interest. As a result of computer development in the past few years, the size of these problems has grown immensely. For small molecules with few electrons, it is possible to use many basis functions and obtain excellent results from the calculations.[21] With larger molecules, the speed of the computer and core limits the number of basis functions that can be used, but significant computations still can be made.

In such computations, by far the most difficult and time-consuming part is the calculation of three- and four-center electron repulsion integrals. A typical such integral can be expressed as

$$\int \int \frac{\varphi_i^A(1)\varphi_j^B(1)\varphi_k^C(2)\varphi_l^D(2)}{r_{12}} \, d\tau(1) \, d\tau(2) \tag{4.1}$$

where $\varphi_i{}^A$, $\varphi_j{}^B$, $\varphi_k{}^C$, and $\varphi_l{}^D$ represent the ith, lth, kth, and lth atomic wave functions centered about points (atoms) A, B, C, and D, and the integration is over the configurational space of electrons one and two. (The electron associated with each function is designated in parentheses.) A commonly used function is the general Slater-type orbital, given by

$$N_i P(r_i) \exp(-\zeta_i r_i) Y_m{}^l(\theta_i, \varphi_i) \tag{4.2}$$

where N_i is the normalization constant, $P(r_i)$ is a simple polynomial in the electron coordinate r_i, $Y_m{}^l(\theta, \varphi)$ is the usual spherical harmonic function, and ζ_i is the (optimized) orbital exponent. The choice of basis set orbitals is quite arbitrary. For instance, we could employ a Gaussian-type orbital.* We have developed a method we believe will greatly reduce the time for calculating such integrals over the arbitrary basis set (by several powers of ten). This method follows from the studies of the anharmonic oscillator.[22] When an extension of the theory is applied to the multicenter integrals of molecular quantum mechanics, the method is as follows.

Let Φ^A be the first members of an infinite complete set of functions centered at A (one atomic center). Likewise, Φ^B, Φ^C, and Φ^D are similar sets of functions centered at points B, C, and D. Form the direct sum of sets Φ^A and Φ^B, and let Φ_i^{AB} be any member of the combined set. That is,

$$\Phi_i^{AB} \; \varepsilon \{ \Phi^A \oplus \Phi^B \} \tag{4.3}$$

where Φ_i^{AB} is any function centered on either atom A or atom B. Let these be the functions for electron 1. Now form the matrices for x, y, z, and S for the set AB. (S is the overlap matrix.)

$$x_{ij}(1) = \int \Phi_i^{AB}(1) x \Phi_j^{AB}(1)\, d\tau \tag{4.4}$$

where, again, Φ_i^{AB} is one of the functions centered on A or B. Likewise,

$$y_{ij}(1) = \int \Phi_i^{AB}(1) y \Phi_j^{AB}(1)\, d\tau \tag{4.5}$$

$$z_{ij}(1) = \int \Phi_i^{AB}(1) z \Phi_j^{AB}(1)\, d\tau \tag{4.6}$$

---

* The main disadvantage in using Gaussian-type functions is the large number that must be employed in a given calculation. See, for instance, E. Clementi, *J. Comp. Phys.* **1**, 223 (1966).

and the overlap matrix

$$S_{ij}(1) = \int \Phi_i^{AB}(1)\Phi_j^{AB}(1)\, d\tau \tag{4.7}$$

The overlap matrix $\mathbf{S}(1)$ is now transformed to a unit matrix in order to make the finite basis set, $\{\Phi^{AB}(1)\}$, orthonormal. The resulting transformation is applied to the $\mathbf{x}(1)$, $\mathbf{y}(1)$, and $\mathbf{z}(1)$ matrices, yielding $\mathbf{x}'(1)$, $\mathbf{y}'(1)$, and $\mathbf{z}'(1)$ matrices. Since x, y, and z all commute, the $\mathbf{x}'(1)$, $\mathbf{y}'(1)$, and $\mathbf{z}'(1)$ matrices may be simultaneously diagonalized, yielding the one-electron transformation T_1. That is,

$$T_1^{-1}\mathbf{x}'(1)T_1 = \boldsymbol{\lambda}_x(1) = |\,\mathbf{x}_i(1)\,|$$

$$T_1^{-1}\mathbf{y}'(1)T_1 = \boldsymbol{\lambda}_y(1) = |\,\mathbf{y}_i(1)\,| \tag{4.8}$$

$$T_1^{-1}\mathbf{z}'(1)T_1 = \boldsymbol{\lambda}_z(1) = |\,\mathbf{z}_i(1)\,|$$

These eigenvalues correspond to a set of spatial points, $\mathbf{r}_i(1) = \{x_i(1), y_i(1), z_i(1)\}$. The simultaneous diagonalizing transformation matrix T_1 is saved. The same type of process is then repeated for electron 2 referenced to atomic centers C and D. This yields a second simultaneous diagonalyzing transformation for electron 2, T_2. That is,

$$T_2^{-1}\mathbf{x}'(2)T_2 = \boldsymbol{\lambda}_x(2) = |\,\mathbf{x}_i(2)\,| \cdots \tag{4.9}$$

and

$$\mathbf{r}_i(2) = \{x_i(2), y_i(2), z_i(2)\} \tag{4.10}$$

The $\mathbf{r}(1)$ and $\mathbf{r}(2)$ matrices may be written in this direct product representation as $\mathbf{r}(1) \otimes \mathbf{1}$ and $\mathbf{1} \otimes \mathbf{r}(2)$ and the transformation that simultaneously diagonalizes $\mathbf{r}(1)$ and $\mathbf{r}(2)$ is then $T_2 \otimes T_1 = T_f$. We then find the distance between each point as

$$\mathbf{r}_{12} = |\,\mathbf{r}_i(1) + R_0 - \mathbf{r}_i(2)\,| \tag{4.11}$$

where R_0 is the vector connecting the centers of the function set Φ^{AB} with those of Φ^{CD}. Now the reciprocal of each distance is formed and an $\mathbf{r}_{12}^{-1}$ matrix is generated. Since $\mathbf{r}_{ij}$ is diagonal, then $\mathbf{r}_{ij}^{-1}$ also will be diagonal with the reciprocal of elements of the r_{ij} matrix. This diagonal matrix is the $\mathbf{r}_{12}^{-1}$ matrix in the direct product representation, where $\mathbf{r}(1)$ and $\mathbf{r}(2)$ are *both* diagonal.* The matrix of *any* function of $\mathbf{r}(1)$ and $\mathbf{r}(2)$ also would

---

* The exact $\mathbf{V}(r_{ij})$ matrix as a function of other matrices ($\mathbf{r}_i(1) = \{x_i(1), y_i(1), z_i(1)\}$, and $\mathbf{r}_i(2) = \{x_i(2), y_i(2), z_i(2)\}$ is defined in terms of Taylor's series in terms of the

be diagonal with

$$f\{\mathbf{r}(1), \mathbf{r}(2)\}_{ik|jl} = f\{[\mathbf{r}(1)]_{ij}, [\mathbf{r}(2)]_{kl}\} \tag{4.12}$$

or

$$f\{\mathbf{r}(1), \mathbf{r}(2)\}_{ik|ik} = f\{[\mathbf{r}(1)]_{ii}, [\mathbf{r}(2)]_{kk}\} \tag{4.13}$$

or

$$\{\mathbf{r}_{12}^{-1}\}_{ik|jl} = \frac{1}{\{|\,[\mathbf{r}(1)]_{ii} - [\mathbf{r}(2)]_{kk}\,|\}} \tag{4.14}$$

The last step is the transformation back to the initial basis set with the inverse of the transformations as

$$\mathbf{T}_f\{\mathbf{r}_{12}^{-1} \text{ (diagonal)}\}\,\mathbf{T}_f^{-1} = \mathbf{E} \tag{4.15}$$

This matrix $\mathbf{E}$ is then the $\mathbf{r}_{12}^{-1}$ matrix in the representation of the original basis functions and is made up of the elements, each of which is one of the desired multicentered integrals.

Since the matrix $\mathbf{r}_{12}^{-1}$ is diagonal, the transformation back to the original representation may be simplified. This may be written as a simple double sum:

$$\int\int \frac{\varphi_i^A(1)\varphi_j^B(1)\varphi_l^C(2)\varphi_k^D(2)}{r_{12}}\,d\tau_1\,d\tau_2 = \sum_m \mathbf{T}_1(i, m)\mathbf{T}_1(j, m)$$
$$\times \sum_n \mathbf{T}_2(k, n)\mathbf{T}_2(l, n)\mathbf{r}_{12}^{-1} \tag{4.16}$$

and it is convenient to carry out the summation for only the particular integrals that are desired. Equation (4.16) yields a means whereby it is not necessary to provide data storage for the direct product transformation, thereby saving much computer storage and time.

It should also be noted here that if the *molecular structure* of the problem is varied such that the *AB* distance and the *CD* distance is held

---

$\mathbf{r}_i(1)$ and $\mathbf{r}_i(2)$ matrices. Since $\mathbf{r}(1)$ and $\mathbf{r}(2)$ are diagonal, all powers of them also will be diagonal and the sum represented by the Taylor's series will simply generate a diagonal matrix with each diagonal element the function of the corresponding diagonal element of the $\mathbf{r}(1)$ and $\mathbf{r}(2)$ matrices. The Taylor's series of expansion must converge at each set of values in which it is used. The $\mathbf{r}_{12}^{-1}$ matrix is a good approximation for the exact $\mathbf{V}(r_{ij})$ matrix at all but very small values of r_{12}. At small values of r_{12}, the actual function $\mathbf{V}(r_{12})$ gives rise to a valid Taylor's series expansion, and no mathematical problems arise at $r_{12} = 0$. These points are avoided because convergence becomes too slow.

constant but all other distances are varied, it is only necessary to repeat the computation starting at Eq. (4.16).

In order to compute most two-, three-, and four-center electron repulsion integrals, it is necessary to "permute" the atomic centers. This "permutation" is equivalent to forming six different direct sum bases [cf. Eq. (4.3)]. The sets of direct sum bases are denoted in brackets and are

I. $\qquad\qquad\{\Phi^{AB} \otimes \Phi^{CD}\}$

to calculate integrals of the type $(12|34)$,

II. $\qquad\qquad\{\Phi^{AC} \otimes \Phi^{BD}\}$

to calculate integrals of the type $(13|24)$, and

III. $\qquad\qquad\{\Phi^{AD} \otimes \Phi^{BC}\}$

to calculate the integrals of the type $(14|23)$,

$$(4.17)$$

where the Φ^{AB}, etc. is the same as Eq. (4.3) and by a set we mean that we are dealing with the direct product of the direct sum basis and hence use the $\{\Phi^{AB} \otimes \Phi^{CD}\}$ notation. The symbol $(IJ|KL)$ has the standard meaning of

$$\int \frac{\varphi^I(1)\varphi^J(1)\varphi^K(2)\varphi^L(2)\,d\tau_1\,d\tau_2}{r_{12}}$$

Set I allows us to calculate the two-, three-, and four-center integrals where Φ^{AB} is a function of electron 1 and Φ^{CD} is a function of electron 2. Or, more specifically, a typical four-center integral for the set of functions $\{\Phi^{AB} \otimes \Phi^{CD}\}$ would be

$$\iint \frac{\varphi_i^A(1)\varphi_j^B(1)\varphi_k^C(2)\varphi_l^D(2)\,d\tau(1)\,d\tau(2)}{r_{12}} \qquad (4.18)$$

In the conventional notation, we could also write this integral symbolically as $(AB|CD)$. Using the set $\{\Phi^{AC} \otimes \Phi^{BD}\}$ yields

$$\iint \frac{\varphi_i^A(1)\varphi_j^C(1)\varphi_k^B(2)\varphi_l^D(2)}{r_{12}}\,d\tau(1)\,d\tau(2) \qquad (4.19)$$

and $\{\Phi^{AD} \otimes \Phi^{BC}\}$ gives

$$\iint \frac{\varphi_i^A(1)\varphi_j^D(1)\varphi_k^B(2)\varphi_l^C(2)\,d\tau(1)\,d\tau(2)}{r_{12}} \qquad (4.20)$$

For Set I, one calculates $|\,\mathbf{x}_i(1)\,|$, $|\,\mathbf{y}_i(1)\,|$, and $|\,\mathbf{z}_i(1)\,|$ for the direct sum Φ^{AB} [cf. Eq. (4.8)]. Then repeat for Φ^{CD} and obtain $|\,\mathbf{x}_i(2)\,|$, $|\,\mathbf{y}_i(2)\,|$, and $|\,\mathbf{z}_i(2)\,|$. As previously outlined, one then calculates the $\mathbf{r}_{12}^{-1}$ matrix and uses Eq. (4.15) to transform back to the original representation. In the most general case, where the atomic functions centered on atoms A, B, C, and D are different, one repeats the process for Set II and finally for Set III.

Also, it would be possible to build up a library of r eigenvalues and transformations for Φ^{AB} sets at various AB distances. Then, when integrals for any possible geometrical configuration are desired it is only necessary to start with Eq. (4.16).

Certain two-electron integrals are omitted in this formalism. They are omitted as a natural consequence of using a one-electron direct sum basis of the form Set I, Set II, or Set III. Use of these three sets omits such two-center, two-electron integrals as

$$\int\!\!\int \frac{\varphi_i^A(1)\varphi_j^A(1)\varphi_k^B(2)\varphi_l^B(2)}{r_{12}}\,d\tau(1)\,d\tau(2) \qquad (4.21)$$

or $(AA|BB)$ and

$$\int\!\!\int \frac{\varphi_i^C(1)\varphi_j^C(1)\varphi_k^D(2)\varphi_l^D(2)}{r_{12}}\,d\tau(1)\,d\tau(2) \qquad (4.22)$$

or $(CC|DD)$.

The sets of functions that would give rise to these two integrals are $\{\Phi^{AB} \otimes \Phi^{AB}\}$ and $\{\Phi^{CD} \otimes \Phi^{CD}\}$, respectively. It is obvious that these sets would yield a null $\mathbf{r}_{12}^{-1}$ matrix. That is, the eigenvalues of electron one and electron two would be identical and R_0 of Eq. (4.11) is zero. The omission of these simple integrals causes no special problems in a calculation of molecular properties since they can be evaluated using debugged well-known expansion technique programs that are in wide circulation.[23]

It should be stressed here that the omission of these simple two-electron integrals is a natural result of using Sets I, II, or III. These sets, however, allow us to calculate the very difficult two-, three-, and four-center integrals.

When an element of the $\mathbf{r}_{12}$ matrix is zero or nearly zero, difficulty is encountered. This situation would arise either by accidental coincidence of $r(1)$ points with $r(2)$ points or within a certain set, such as those which would lead to the two-center integrals. This difficulty does not arise as a result of a discontinuity in r_{12}^{-1} as $r_{12} \rightarrow 0$. The reason is that we are not really expanding a $1/r_{12}$ function but rather the potential energy of two electrons separated by a distance r_{12}. In the region where r_{12} does not approach zero, this potential is the r_{12}^{-1} matrix as we have used it. At very small r_{12} distances, the true two-electron potential $V(r_{12})$ does not become large without limit. Instead, it remains finite and is given by whatever short-range forces are operating.

In our expansion, then, the zero values of r_{12} do not affect the validity of the treatment, but they do slow up the convergence of our results to the extent that we will always avoid calculating them. That is, we shall avoid calculating integrals that come from blocks in which an $r(1)$ point is very close to an $r(2)$ point, giving rise to a zero or nearly zero r_{12} value.

If 40 basis functions are used at each of the 4 centers, it would be necessary to diagonalize several 80×80 matrices (4 to 8, depending upon whether the distance between A and B was equal to that between C and D). The next step is the computation of 80×80 or 6400 distances between the two sets of 80 points and to perform the reverse transformation using the two 80×80 transformations and the 6400 distances. These are very large matrices, but the summations in Eq. (4.16) are very fast and the total summing process would result in 6400×6400 or 40,960,000 integrals.

Since a finite number of basis functions are employed in a particular computation, instead of the complete infinite set, errors may be introduced into the calculated matrices. As shown in Harris et al.[22] the errors may be eliminated in all elements of interest for the one-dimensional case by increasing the size of the basis set until no significant change occurs. The eigenvalues and eigenvectors of a matrix $\mathbf{M}$ are a function of the dimension of $\mathbf{M}$. Therefore a finite basis set yields different eigenvectors and eigenvalues than would an infinite set. The eigenvectors constitute only a diagonalyzing transformation for $\mathbf{M}$. Any effects that the difference between an infinite and finite matrix transformation have on the calculation of the r_{12}^{-1} integrals are canceled by the inverse transformation of Eq. (4.15). But since we are dealing with a series expansion of a function, $\{r_{12}^{-1} \text{ (diagonal)}\}$ in terms of $n \times n$ matrices, we shall use the first 5000×5000 values of the $\mathbf{E}$ matrix of Eq. (4.15) rather than 6400×6400 values generated with 40 basis functions on each center in order to avoid expansion errors.

This method yields a far shorter running time than that which would be required to calculate these integrals over the six electronic coordinates by numerical integrations. At present, representative computational times for four-center integrals are of the order of 5 milliseconds for four s-type functions of arbitrary geometry (convulution method of Hagstrom) to 50 milliseconds for four f-type functions in a *linear* array (method of McLean and Yoshimini). In arbitrary geometry, the latter figure increases by an

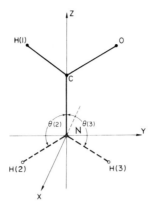

Fig. 1. Position of formamide, $HCONH_2$, in Cartesian coordinate system. H(1), O, C, and N in the yz plane.

order of magnitude.[21] If the method works as well as we anticipate and as well as it did for the anharmonic oscillator,[22] it will then be possible to perform far better calculations of the quantum mechanics of the chemical bonding and of molecular structure than can be done presently.

V. COMPUTATIONAL RESULTS

In this section, we present some brief selected results of the quantum mechanically computed barrier to hindered internal rotation of the form-amide molecule ($HCONH_2$).[10] In particular, the computation was initiated in order to compute the NH_2 group barrier to internal rotation about the C—N bond and other related properties of interest. Figures 1 and 2 define

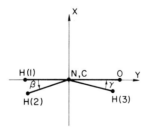

Fig. 2. β and γ are the dihedral angles between the yz plane and the planes H(2)NC and H(3)NC, respectively.

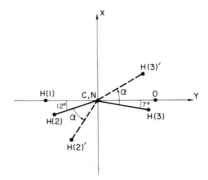

Fig. 3. Definition of angle of rotation α of NH_2
group in formamide around the NC bond (z
axis). $H(2) \rightarrow H(2)'$ and $H(3) \rightarrow H(3)'$.

the Cartesian coordinate system, Fig. 3, the angle of hindered rotation, and
Fig. 4 shows the qualitative features of the computed potential energy curve
for the internal rotation problem. The quantitative results were in satisfactory agreement with experiment and may be found in reference 10 along
with other computational details. Forty-four nuclear configurations were
studied. If each of these calculations were performed from scratch, we would
need to compute roughly: $44 \times (56)^4/8$ integrals of the form given by
Eqs. (3.4 and 3.5).

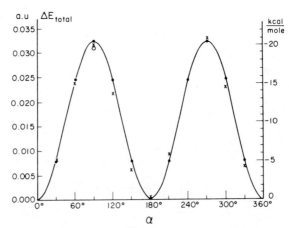

Fig. 4. *Ab initio* potential curve for internal rotation in
formamide around CN bond; angle of rotation α (see
Fig. 2). Unconnected points are explained in Christensen
et al.[10]

References

1. R. S. Mulliken, What are the electrons really doing in molecules? *The Vortex*, Cal. Sec. Am. Chem. Soc. (Spring 1960).

2. C. A. Coulson, *Valence*, Oxford University Press, London, 2nd ed., 1961.

3. J. H. Van Vleck and A. Sherman, The quantum theory of valence, *Revs. Mod. Phys.* **7**, 167–228 (1935).

4. R. G. Parr and F. O. Ellison, The quantum theory of valence, *Ann. Rev. Phys. Chem.* **6**, 171–192 (1955). Also see other reviews on quantum theory in Vols. 1–12, *Ann. Rev. Phys. Chem.*

5. G. G. Hall, Application of quantum mechanics in theoretical chemistry, *Repts. Progr. Phys.* **32**, 1–32 (1959).

6. C. C. J. Roothaan and R. S. Mulliken, Broken bottlenecks and the future of molecular quantum mechanics, *Proc. Natl. Acad. Sci. U.S.* **45**, 394–398 (1959).

7. S. F. Boys and G. B. Cook, Mathematical problems in the complete predictions of chemical phenomena, *Rev. Mod. Phys.* **32**, 285–295 (1960).

8. B. Bak, E. Clementi, and R. N. Kortzeborn, Structure, vibrational spectra, dipole moment, and stability of gaseous LiCN and LiNC, *J. Chem. Phys.* **52**, 764–772 (1970).

9. P. N. Noble and R. N. Kortzeborn, LCAO-MO-SCF studies of HF_2^- and the related unstable systems HF_2^0 and HeF_2, *J. Chem. Phys.* **53**, 5375–5387 (1970).

10. D. H. Christensen, R. N. Kortzeborn, B. Bak, and J. J. Led, Results of *ab initio* calculations on formamide, *J. Chem. Phys.* **53**, 3912–3922 (1970).

11. J. C. Slater, Solid-state and molecular theory group, *Quart. Progr. Rept. No. 43* (January 15, 1962).

12. W. Kauzmann, *Quantum Chemistry*, Academic Press, New York, 1957.

13. L. Pauling and E. B. Wilson, *Introduction to Quantum Mechanics*, McGraw-Hill Book Co., Inc., New York and London, 1935.

14. H. Eyring, J. Walter, and G. Kimball, *Quantum Chemistry*, John Wiley and Sons, Inc., New York and London, 1960.

15. L. Landau and E. Lifshitz, *Quantum Mechanics: Non-Relativistic Theory*, Addison-Wesley Publishing Co., Reading, Massachusetts, 1958.

16. L. Schiff, *Quantum Mechanics*, McGraw-Hill Book Co., Inc., New York, 1955.

17. J. Powell and B. Crasemann, *Quantum Mechanics*, Addison-Wesley Publishing Co., Reading, Massachusetts, 1961.

18. C. H. Wilcox, ed., *Perturbation Theory and its Applications in Quantum Mechanics*, John Wiley and Sons, Inc., New York, 1965.

19. V. Fock, Näherungsmethode zur Lösung des quantummechanischen Mehrkörperproblems, *Z. Phys.* **61**, 126 (1930).

20. A. C. Wahl, P. E. Cade, and C. C. J. Roothaan, The study of 2-center integrals useful in calculations on molecular structure. V. General methods for diatomic integrals applicable to digital computers, *J. Chem. Phys.* **41**, 2578 (1964).

21. C. J. Roothaan and P. S. Bagus, *Methods in Computational Physics*, Academic Press, Inc., New York, 1963, Vol. II.

22. D. O. Harris, G. Engerholm, and W. D. Gwinn, *J. Chem. Phys.* **43**, 1515, 1965. This paper applies a well established mathematical method in a new way, which greatly facilitates the computer solution of many problems in quantum mechanics.

23. F. J. Corbato and A. C. Switendeck, *Methods in Computational Physics*, Academic Press, Inc., New York, 1963, Vol. II.

24. A. D. McLean, IBM Research Center, San Jose, California (private communication).

25. R. N. Kortzeborn, *The Virtual Computer and Its Importance to Scientists*, IBM Data Processing Division, Palo Alto Scientific Center report in preparation.

APPENDIX:
Fortran Code

```
C       THIS IS THE MAIN DECK THAT CALLS ALL OF THE SUBROUTINES....
        IMPLICIT REAL*8(A-H,O-Z)
        DIMENSION A1(72,72),A2(72,72),A3(72,72),AL(6,72),Q(15),
       2FFACT(1),FACT(31),L(72),M(72),NZ(72),ZETA(72)
        CALL FSPIE
         REWIND 2
        REWIND 3
        FFACT(1) = 1.0
        DO 2001 I=1,2
        II = I-1
        FACT(II) = 1.0
 2001 CONTINUE
        DO 2000 I=2,30
        TEMP = I
 2000 FACT(I) = FACT(I-1)*TEMP
C       KK = 1 FOR THE AB SET.....KK=4 FOR THE CD SET.....
C       KREP =1 IF THE SETS ARE THE SAME AND 2 IF DIFFERENT.....
        READ (5,4) KREP,NDIM
        WRITE (6,15) KREP,NDIM
   4    FORMAT (215)
        DO 3 KI = 1,KREP
        REWIND 2
        REWIND 3
        READ (5,30) (Q(I),I=1,15)
   30   FORMAT (6F12.8)
        READ (5,2) KK,ID,DISTAB
        WRITE (6,2) KK,ID,DISTAB
C        ID IS NOT USED,KK=1 FOR SET B AND KK=4 FOR SET CD
   2    FORMAT (215,F20.8)
        REQ = DISTAB
C       READ ORBITAL INPUT DATA
C..................................................................
C       THIS PART OF THE PROGRAM GENERATES ALL THE STOS FOR THE
C       CALCULATION AND ORDERS THEM A LIST.  THE LENGTH OF THIS LIST
C       IS EQUAL TO MAX.........
C       THE INPUT CARD
C       THE INPUT CARDS CONTAIN ONLY THE L,N AND ZETA QUANTITIES FOR
C       EACH STO.  THE PROGRAM THEN GENERATES ORBITALS FOR ALL POSITIVE
C       AND NEGATIVE M VALUES CORRESPONDING TO EACH L VALUE..
C       THESE ARE THE QUANTITIES WHICH ARE PUT INTO THE LIST OF LENGTH
C       MAX.
C       RULES FOR READING THE ORBITALS ARE AS FOLOWS......
C.....READ ALL S TYPE FUNCTIONS ON CENTER A FIRST.
C       THE NUMBER OF THESE IS KSORB...
C       FOLLOW THIS BY THE NUMBER OF NON-STYPE FUNCTIONS ON CENTER A....
C       THE NUMBER OF THESE IS KOTHER.....
C       NOW REPEAT(WITH S TYPE FUNCTIONS FIRST) FOR CENTER B........
C       THIS PROCEDURE IS FOLLOWED FOR EACH DIRECT SUM SET............
C..................................................................
        KC = 0
        KSET = 0
   45   READ (5,15) KSORB,KOTHER
        WRITE (6,15) KSORB,KOTHER
   15   FORMAT (215)
        LC = KC + 1 - KSET
        LCC = KSORB + LC - 1
        WRITE (6,15) LC,LCC
        IF(KSORB) 46,46,47
   47   DO 42 KC = LC,LCC
        WRITE (6,15) KC
C       READ IN THE S TYPE STOS.
        READ (5,1) L(KC),NZ(KC),ZETA(KC)
        M(KC) = 0
   1    FORMAT (I3,6X,I3,F10.8)
   42   CONTINUE
   46   KC = LCC + 1
        IF(KOTHER) 48,48,49
   49   DO 41 K=1,KOTHER
C       READ IN THE NON-S TYPE STOS...
        READ (5,1) L(KC),NZ(KC),ZETA(KC)
        WRITE (6,15) KOTHER,KC
        KA = 0
        KQ = L(KC)
        KKN = NZ(KC)
        AKZ = ZETA(KC)
        KP = 2*KQ + 1
        DO 40 KB = 1,KP
        M(KC)=KQ-KA
        L(KC) = KQ
        NZ(KC) = KKN
        ZETA(KC) = AKZ
        KA = KA + 1
        KC = KC + 1
   40   CONTINUE
```

```
41    CONTINUE
48    IF(KSET) 44,44,43
44    MAXA=KC-1
      KSET=1
      GO TO 45
43    CONTINUE
      MAXB = MAX
      MAX = KC - 1
      WRITE (6,15) MAX,KC
      WRITE (6,260)
260   FORMAT (1H1,39H THE INPUT STOS ARE(L,M,N,ZETA,ORB NO.))
      WRITE (6,1015) (L(I),M(I),NZ(I),ZETA(I),I,I=1,MAX)
1015  FORMAT (3I5,F20.8,I5)
      WRITE (6,261) MAX,NDIM
261   FORMAT (19H MAX AND NDIM ARE =,2I5)
C        SUBROUTINE TWO IS NOW CALLED.  IF KXINT=KZINT=0, THE OVERLAP
C        MATRIX IS CALCULATED.  AFTER COMING OUT OF TWO, THE OVERLAP
C        MATRIX IS WRITTEN ON LOGICAL TAPE 2.
      KXINT=0
      KZINT=0
      CALL TWO(ZETA,NZ,L,M,REQ,FACT,NDIM,MAX,MAXA,KXINT,KZINT,A3)
      WRITE (2) A3
C        KXINT = 1 AND KZINT = 0 GOING INTO TWO CAUSES THE X AND Y MATRICES
C        TO BE CALCULATED.
      KXINT=1
      CALL TWO (ZETA,NZ,L,M,REQ,FACT,NDIM,MAX,MAXA,KXINT,KZINT,A1)
C.....THIS CALCULATES THE MATRIX 2*X.....WHICH IS EQUAL TO 2*Y
C.....Y IS PURE IMAGINARY SO WE  DONT WRITE IT ON TAPE...
C        KZINT = 1 AND KXINT = 0 CAUSES THE Z MATRIX TO BE CALCULATED.
      KXINT = 0
      KZINT = 1
      CALL TWO(ZETA,NZ,L,M,REQ,FACT,NDIM,MAX,MAXA,KXINT,KZINT,A2)
      DO 500 I=1,MAX
      DO 501 J=1,MAX
      A1(I,J) = A1(I,J)/2.0
      IF(I-MAXA) 501,501,502
502   IF(J-MAXA) 501,501,503
503   A2(I,J) = A2(I,J) + DISTAB*A3(I,J)
501   CONTINUE
500   CONTINUE
      WRITE(2) A1
      WRITE (2) A2
      IA = KK + KREP - 3
      REWIND 2
      CALL ONE(A1,A2,A3,Q,NDIM,MAX,KK,AL,IA,MAXB)
3     CONTINUE
      CALL DIRECT(MAX,NDIM,A1,A2,A3,AL,MAXB)
      STOP
      END


      SUBROUTINE TWO(ZETA, NZ,L,M,REQ,FACT,NDIM,MAX,MAXA,KXINT,KZINT,A1)
      IMPLICIT REAL*8(A-H,O-Z)
      DIMENSION ZETA(NDIM),NZ(NDIM),L(NDIM),M(NDIM),D(2,400),
     1                    A1(NDIM,NDIM),FFACT(1),FACT(31)
      FFACT(1) = 1.0
C        THIS SUBROUTINE CALCULATES THE ONE AND TWO CENTER INTEGRALS NEEDED
C        FOR EACH SET.  THEY ARE THE OVERLAP, THE X AND Z MATRICES,....
C        THE TWO CENTER INTEGRALS ARE CALCULATED IN TERMS OF PROLATE
C        SPHERICAL COORDINATES DEFINED IN TERMS OF THE FOLLOWING SPHERICAL
C        COORDINATES.......
C        XI OR X =(R(A)+R(B))/R
C        ETA OR Y = (R(A)-R(B))/R
C        PHI = PHI(A) = PHI(B)
C        REFERENCE IS J. CHEM. PHYS. 41, 2578(1964).....
C        REQ IS THE INTERNUCLEAR DISTANCE AB OR REQ IS THE INTERNUCLEAR
C        DISTANCE CD.  (SECOND PASS, IF CD IS NE TO AB)....
C        ONE CENTER INTEGRALS ARE CALCUATED WITH THE ORIGIN AT
C        THEIR CENTER...
C        THE AB INTEGRALS ARE CALCULATED WITH THE ORIGIN AT CENTER A...
C        SINCE Y(L,M) = ((-1)**M)*Y(L,-M), NEGATIVE VALUES OF M ARE NOT
C        CALCULATED, BUT ARE GIVEN IN TERMS OF THE CORRESPONDING INTEGRALS
C        WITH M POSITIVE.  THIS REQUIRES THE BASIS SET TO CONTAIN ALL
C        POSSIBLE M VALUES.  , STARTING WITH +L AND GOING TO -L
C        KXINT AND KZINT ARE FLAGS, EITHER 0 OR 1, AND AND INDICATE
C        THAT THE CALCULATIONS ARE TO RETURN THE X OR THE Z MATRIX
C        INSTEAD OF THE OVERLAP MATRIX....
C        THE CONVENTION THAT MUST BE USED IS ....READ IN ORBITALS ON CENTER
C        A, THEN ON CENTER B FOR SET AB...
C        READ IN ORBITALS ON CENTER C, THE ON D FOR SET CD
C        THIS READING IS DONE IN THE MAIN PROGRAM AND MUST BE ADHERED TO..
      KXZ = KXINT + KZINT
      DO 101 I=1,MAX
```

```
        DO 102 J=1,I
C       CHECK M ORTHOGONALITY HERE...NOTE THAT THE CALCULATION HAS BEEN
C       TRIANGULATED DUE TO SYMMETRY...
        SGNI = 1.0
        IP = I
        JP = J
        IF(IABS(M(I)-M(J)).NE.KXINT) GO TO 1011
        IF((KXINT.EQ.0).AND.(KZINT.EQ.0).AND.(I.EQ.J)) GO TO 103
        IF(KXINT-1) 301,300,301
  300   SGNI = -1.0
  301   IF(M(I)) 302,303,303
  302   IP = I+2*M(I)
        IF(M(J))304,305,305
  303   IF(M(J)) 304,1,1
  304   JP = J+2*M(J)
  305   A1(I,J) = SGNI*A1(IP,JP)
        GO TO 53
 1011   A1(I,J) = 0.
        GO TO 53
C       THE SIGNED M ON A AND MA ON B MUST BE USED IN THE ORTHOGONALITY
C       CHECK. (KRONECKER DELTA).....
C       IN THE DOUBLE INTEGRATION OVER WIGGLE AND ETA, ONLY POSITIVE M
C       VALUES NEED BE CONSIDERED.  WE JUST OBTAIN THE INTEGRALS FOR
C       M ON A AND M ON B BEING POSITIVE AND MULTIPLY THE END INTEGRAL
C       BY THE APPROPRIATE FACTOR OF -1 FOR CASES WHERE M ON A OR M ON B
C       IS NEGATIVE.....
C       IF CALCULATING THE OVERLAP MATRIX, SET THE DIAGONAL ELEMENTS =1
    1   MM = M(I)
    3   IF(I-MAXA)402,402,401
  401   IF(J-MAXA)403,403,402
C       GO TO 402 IF THIS IS A ONE CENTER INTEGRAL
C       GO TO 403 IF THIS IS A TWO CENTER INTEGRAL
C*****CHECK ALL OF THE TRIANGLE CONDITONS FOR THE X AND Z MATRICES....
  402   IF((IABS(L(I)-L(J))).NE.KXZ) GO TO 1011
        Z1 = ZETA(I)
        Z2=ZETA(J)
        ZZ=Z1+Z2
        Z3=2.*Z1
        Z4=2.*Z2
        N1=NZ(I)
        N2=NZ(J)
        N3=2*N1
        N4=2*N2
        N5 = N1+N2+KZINT+KXINT
        SQ = DSQRT((Z3*Z4)/(FACT(N3)*FACT(N4)))
        A1(I,J)=SQ*(Z3**N1)*(Z4**N2)*FACT(N5)/(ZZ**(N5+1))
        FL = L(I)
        LL = FL
        FM = MM
        SIGN = +1.0
        IF(KXINT-KZINT)421,53,422
C       GO TO 421 IF THIS IS A ONE CENTER Z MATRIX ELEMENT
C       GO TO 422 IF THIS IS A ONE CENTER X MATRIX ELEMENT
  421   IF(LL-L(J))423,1011,425
  423   FFL = (FL-FM+1.)*(FL+FM+1.)/((2.*FL+1.)*(2.*FL + 3.))
        GO TO 436
  425   FFL = (FL-FM)*(FL+FM)/((2.*FL-1.)*(2.*FL+1.))
        GO TO 436
  422   IF(LL-L(J))433,1011,434
  433   IF(MM-M(J))427,1011,428
  427   FFL = (FL-FM+2.)*(FL+FM+1.)/((2.*FL+1.)*(2.*FL+3.))
        SIGN = -1.0
        GO TO 436
  428   FFL = (FL-FM+1.)*(FL-FM+2.)/((2.*FL+1.)*(2.*FL+3.))
        GO TO 436
  434   IF(MM-M(J))437,1011,438
  437   FFL = (FL-FM)*(FL-FM-1.)/((2.*FL+1.)*(2.*FL-1.))
        GO TO 436
  438   FFL = (FL+FM-1.)*(FL+FM)/((2.*FL-1.)*(2.*FL+1.))
        SIGN = -1.0
  436   A1(I,J) = A1(I,J)*DSQRT(FFL)*SIGN
        GO TO 53
  403   RHO = (ZETA(I)+ZETA(J))*REQ/2.0
        TAURHO = (ZETA(I) - ZETA(J))*REQ/2.0
        A1(I,J) = 0.0
C       GET THE VA AND VB COEFFICIENTS.
        N1 = 2*NZ(I)
        VA = FACT(N1)*(ZETA(I)*REQ)**(-N1-1)
        N2 = 2*NZ(J)
        VB = FACT(N2)*(ZETA(J)*REQ)**(-N2-1)
        KR = (L(I) - IABS(M(I)))/2 + 1
        KS = (L(J) - IABS(M(J)))/2 + 1
C       THE SUMS OVER KR AND KS ARE NOT OBVIOUS.  THEY GO FROM 0 TO KR
C       AND 0 TO KS..........BUT..........
```

```
C      THEY GO IN INTEGER STEPS.  IF KR OR KS TURNS OUT TO BE SOME
C      FRACTION SUCH AS ONE-HALF, THE SUM IS ONLY PERFORMED FOR ZERO.
C      IF KR OR KS IS 3/2, THE SUM IS DONE FOR ZERO AND ONE,ETC...
C      SO ONE CAN USE FIXED POINT TRUNCATION FOR THIS PURPOSE.
C      ONE SPECIAL CASE NEEDS TO BE TAKEN INTO ACCOUNT.
C      THAT IS WHEN KR OR KS = 0.  THIS CAN HAPPEN WHEN KR OR KS = 1/2
C      ALSO BECAUSE THE MACHINE WILL ROUND THEM DOWN TO ZERO.
C      WHEN THIS HAPPENS, SET KR AND KS = 1.  THEN KI = 0 AND KJ = 0
C      AND THE WR AND WS COEFFICIENTS WILL BE PROPERLY EVALUATED FOR 0
C      ONLY.
C      SUM OVER THE KI INDEX (0 TO KR) AND THE KJ INDEX (0 TO KS)
       TEM=0.
       WR = 0.0
       DO 700 KKI = 1,KR
       KI = KKI - 1
C      SET UP THE W COEFFICIENTS FOR EACH KI AND KJ
       ONE = (2*L(I) + 1)
       ONEA = ONE/2.0
       N3 = L(I) - M(I)
       N4 = L(I) + M(I)
       TWO = FACT(N3)/FACT(N4)
       PHA = (-1)**(M(I) + KI)
       N5 = (2*L(I) - 2*KI)
       THREE = FACT(N5)
       FOUR = 2**L(I)
       N6 = (L(I) - M(I) - 2*KI)
       FIVE = FACT(N6)
       N7 = (L(I) - KI)
       SIX = FACT(N7)
       SEVEN = FACT(KI)
       WR =(DSQRT(ONEA*TWO)*PHA*THREE)/(FOUR*FIVE*SIX*SEVEN)
       WS = 0.0
       DO 699 KKJ = 1,KS
       KJ = KKJ - 1
       ONE = (2*L(J) + 1)
       ONEA = ONE/2.0
       M1 = L(J) - M(J)
       M2 = L(J) + M(J)
       TWO = FACT(M1)/FACT(M2)
       PHA = (-1)**(M(J) + KJ)
       M3 = (2*L(J) - 2*KJ)
       THREE = FACT(M3)
 1111  FOUR=2**L(J)
       M4 = (L(J) - M(J) - 2*KJ)
       FIVE = FACT(M4)
       M5 = (L(J) - KJ)
       SIX = FACT(M5)
       SEVEN = FACT(KJ)
       WS =(DSQRT(ONEA*TWO)*PHA*THREE)/(FOUR*FIVE*SIX*SEVEN)
C      SET UP INDECES FOR THE L INTEGRAL...
       IF(KXZ) 411,411,412
  412  WR=WR*REQ/2.0
  411  KA=NZ(I)-L(I)+2*KI+1
       KB = NZ(J) - L(J) + 2*KJ + 1
       KG = L(I) -IABS(M(I)) - 2*KI + 1+KZINT
       KD = L(J) -IABS(M(J)) - 2*KJ + 1
       KF = 1 + (IABS(M(I)) + IABS(M(J)) + KXINT )/2
C      THE ABOVE CONSTANTS ARE EQUAL TO THE NUMBER OF TERMS IN THE
C      BINOMIAL EXPANSIONS AND ARE 1 GREATER THAN THE POWER
C*****CHECK TO SEE IF THE POWER OF X AND Y ARE ZERO...............
       IC = KA+KB+KG+KD+KF-5
       IF (IC) 240,240,241
  240  TERM = AAUX(RHO,0)*BAUY(TAURHO,0,FACT)
       GO TO 212
  241  DO 200 N=1,400
       D(1,N)=0.
       D(2,N)=0.
  200  CONTINUE
       D(1,1)=1.
       I1=1
       I2=2
       JA=KA-1
       IF(KA-1) 110,110,112
C      THIS IS (X+Y)**KA
  112  KPX=1
       JPX=0
       JPY=1
       CALL PMUL(KA,JA,KPX,JPX,JPY,D,FACT,I1,I2)
  110  JA=JA+KB-1
       IF(KB-1) 114,114,116
C      THIS IS (X-Y)**KB
  116  KPX=1
       JPX=0
       JPY=-1
```

```
            CALL PMUL(KB,JA,KPX,JPX,JPY,D,FACT,I1,I2)
    114     JA=JA+2*KG-2
            IF(KG-1)117,117,119
    119     KPX=0
            JPX=1
            JPY=1
            CALL PMUL(KG,JA,KPX,JPX,JPY,D,FACT,I1,I2)
    117     JA=JA+2*KD-2
            IF(KD-1) 120,120,122
C           THIS IS (1-XY)**KD
    122     KPX=0
            JPX=1
            JPY=-1
            CALL PMUL(KD,JA,KPX,JPX,JPY,D,FACT,I1,I2)
    120     JA=JA+2*KF-2
            IF(KF-1) 123,123,125
C     THIS IS (X**2-1)**KF
    125     KPX=-2
            JPX=0
            JPY=0
            CALL PMUL(KF,JA,KPX,JPX,JPY,D,FACT,I1,I2)
            JA=JA+2*KF-2
C           THIS IS (1-Y**2)**KF
            KPX=0
            JPX=0
            JPY=-2
            CALL PMUL(KF,JA,KPX,JPX,JPY,D,FACT,I1,I2)
    123     CONTINUE
C           NOW MULTIPLY THE FINAL COEFFICIENT BY THE APPROPRIATE INTEGRAL
    51      TERM = 0.0
            JA=JA+1
            DO 211 N=1,JA
            IT=N*(N-1)
            IT=IT/2
            DO 210 JI=1,N
            ITT=IT+JI
            DD=D(I1,ITT)
            DDL=1.E-8
            IF(DABS(DD)-DDL)210,210,213
    213     JQ=JI-1
            IQ=N-JI
            AX = AAUX(RHO,IQ)
            BY=BAUY(TAURHO,JQ,FACT)
            TERM=TERM+DD*AX*BY
    210     CONTINUE
    211     CONTINUE
    212     TEM=TEM+TERM*WR*WS
    699     CONTINUE
    700     CONTINUE
            A1(I,J) = TEM*(1.0/DSQRT(VA*VB))
            GO TO 53
    103     A1(I,J) = 1.0
C           NOW REFLECT THROUGH THE MAIN DIAGONAL
    53      A1(J,I) = A1(I,J)
    102      CONTINUE
    101     CONTINUE
            IF(KXINT) 52,52,6000
    6000    CONTINUE
            DO 6001 I=1,MAX
            DO 6001 J=1,MAX
            A1(I,J) = -1.0*A1(I,J)
    6001    CONTINUE
    52      RETURN
            END
```

Chapter 6

Numerical Simulation of Weather

Harwood G. Kolsky

IBM Scientific Center
Palo Alto, California

I. PHYSICAL PHENOMENA

A. Introduction

Even in this age of scientific superlatives it is hard to find a field more far-reaching and with more interesting problems and more difficulties than that of numerical weather prediction. The associated atmosphere physics behind it is worldwide. A large number of physical disciplines interact with each other in a most complex way. Fluid dynamics, which describes the major motions of the atmosphere and oceans, is considered classical physics. However, the energy sources and frictional forces that must be included introduce quantum mechanics and diffusion theory as well as numerical analysis.

Although one cannot underestimate the importance of physical theory and numerical methods, the real history of numerical weather prediction has been essentially tied to that of the speed and capacity of the computers available. Richardson's attempt[1] about 1920 to perform numerical weather calculations starting from the primitive equations of hydrodynamics is well known. It is less well known that he proposed tying the calculations into an operational network. The magnitude of his effort and the difficulties he encountered were enough to discourage meteorologists from serious work in this area for more than twenty years. However, reading his account now is

173

an interesting experience because it sounds surprisingly modern. He pointed out some of the main pitfalls to be avoided by later workers.[36] One would have to make relatively few serious changes in Richardson's book to bring it out as a modern treatise in the field. This was the spirit in which Kasahara and Washington[19] developed their recent model at the National Center of Atmospheric Research at Boulder, Colorado.

The purpose of this section is to survey the basic physical phenomena concerning the atmosphere and some of the numerical and computer design problems that arise in attempting to model it. The emphasis is on general circulation research problems rather than on existing operational methods. Our feeling is that these research problems are the ones which stretch the state of the art, place the heaviest demands on computers, and point the way to the future.

B. Magnitude and Range of the Problem

In discussing the requirements for improved weather forecasting, Phillips[2] offers the following comment:

> Unfortunately, faster computing machines are not the only requirements for improved weather predictions. The basic physical equations, which are nonlinear, presuppose an extremely detailed knowledge of the state of the atmosphere at the beginning of the forecast. For example, the viscosity term in the Navier–Stokes hydrodynamic equation is of fundamental importance because it is ultimately responsible for the frictional dissipation of kinetic energy in the atmosphere. However, it can perform this vital function in a numerical calculation only if the latter includes motion on scales as small as a millimeter. Analogous difficulties appear in other equations, especially those describing condensation of water vapor and precipitation (where the fundamental physical laws apply to individual raindrops) and radiation efforts (where the molecular spectra are extremely complicated). The most important weather phenomena, on the other hand, have horizontal scales of 10^5 to 10^7 meters, and experience has shown that it is necessary to consider conditions over almost an entire hemisphere to predict the weather several days in advance. It is obviously impractical to allow for the scale ratio of 10^{10} in any conceivable computation scheme.

The energy involved in the weather is much larger than is usually realized. All human electrical output is about equal in power to a big thunderstorm. A large storm system can be 100 times greater.

C. Scaling Approximations

To deal with these difficulties of scale many kinds of approximations in the field of dynamic meteorology have been developed. They are all concerned with including or omitting certain physical quantities from the model. As a consequence, the solutions of such approximate, discrete systems do not describe all physical phenomena associated with the complete system of differential equations. Such approximations are said to "filter out" entire ranges of phenomena, e.g., sound waves, which should not have an affect on the answers of interest such as large-scale cyclonic motion.

In the scale-analysis approach, it is assumed that all dependent variables —such as the velocity—are characterized by a "well-defined rate of variation," i.e., a scale, in space and time. In particular, it is assumed that a partial derivative of a quantity will have an order of magnitude at most equal to the magnitude of the quantity divided by the appropriate scale length. This scale-analysis method has the advantage of maintaining a relatively clear and unambiguous relationship between the "true" variables in the atmosphere and the variables in the simplified equation. It still requires some physical intuition, however, since definite statements about the order of magnitude of various quantities are necessary.

Table I. Scales of Motion Associated with Typical Atmospheric Phenomena[19]

Classification	Scale, km	Typical phenomena
Large scale	10,000	Very long waves
	5,000	Cyclones and anticyclones
Medium scale	1,000	Frontal cyclones
	500	Tropical cyclones
Mesoscale	100	Local severe storms, squall lines
	50	Hailstorms, thunderstorms
Small scale	10	Cumulonimbi
	5	Cumuli
	1	Tornados, waterspouts
Microscale	0.5, 0.1	Dust devils, thermals
	0.05, 0.01	Transport of the heat, momentum, and water vapor

General circulation calculations fit into the middle range of physical phenomena. (Their characteristic time scale is from 10^3 to 10^5 seconds, and their characteristic length scale is from 10^5 to 10^7 meters.) These are motions in which the Coriolis force from the earth's rotation is important.

Fortunately, motions on this large scale are not only responsible for most day-to-day weather changes (and are therefore worth forecasting), but it seems that their behavior can be predicted satisfactorily over periods of several days without too much detailed consideration of the unknown smaller-scale phenomena. Table I shows that "mesoscale" and "small scale" phenomena in this case include individual thunderstorms, tornados, and even hurricanes. Except for special warning networks, they fall outside the regular observations and realistic computing models.

D. Fundamental Conservation Equations

1. Physical Basis

The physical and mathematical basis of all methods of dynamic weather prediction lies in the principles of conservation of momentum, mass, and energy. Applied to the quasicontinuous statistical motion of an assemblage of liquid or gas molecules (through the methods of kinetics theory and statistical mechanics), these fundamental principles are expressed mathematically in Newton's equations of motion for a continuous medium, the equation of continuity (for mass conservation), and the thermodynamic energy equation. So far as known, these equations are universal in that they evidently apply to all fluids in normal ranges of pressure, temperature, and velocity, without regard to composition, container, or state of motion.

The basic equations contain terms that describe such physical entities as heat influx, water vapor influx, and frictional forces, which after discretization and scaling assumptions are no longer implicitly described by the system. Such phenomena are said to be "parameterized" in the model. Much of the work done by researchers in the field can be characterized as an attempt to remove parameterization or to find more efficient and effective numerical methods of parameterization.

Fortunately, by far the most rapid accelerations of the earth relative to the fixed stars are those due to its rotation around a polar axis whose absolute direction is fixed. This permits us to regard the earth as a body in pure rotation relative to an inertial frame traveling along with the earth.

2. *Equations of Motion*

Following the development used by P. D. Thompson:[4] If we consider any vector field A, which is a function of time and space, its total derivative with respect to time may be expressed as follows:

$$\frac{dA}{dt} = \frac{d'A}{dt} + \Omega \times A$$

where d'/dt denotes total differentiation relative to the rotating frame. This equation states that the difference between the total derivatives of any vector field A, measured relative to the inertial and rotating frames, is equal to the cross product (or vector product) of A and the angular velocity Ω of the rotating frame—in our case, the earth.

In particular, we can identify A with the position vector r, which is defined as

$$r = xi + yj + zk = x'i' + y'j' + z'k'$$

and with the velocity vector $U = dr/dt$ the equations become

$$\frac{dr}{dt} = \frac{d'r}{dt} + \Omega \times r$$

which is the same as $U = U' + \Omega \times r$ and $dU/dt = d'U/dt + \Omega \times U$.

It is now possible to derive Newton's second law, as applied to motions relative to the earth. Equating the absolute acceleration to the resultant of all real forces (per unit mass), we have

$$\frac{dU}{dt} = \frac{d'U'}{dt} + 2\Omega \times U' - \Omega^2 R = F + g^* + P$$

where F symbolizes the force of internal viscosity per unit mass, g^* is the gravitational force, and P is the force due to spatial variations of atmospheric pressure. Since the real forces are the same, whether referred to the inertial frame or the rotating frame, the forces F, g^*, and P may be referred to the rotating frame.

Owing to the fact that the gravitational and centrifugal forces are functions of position alone and do not depend on motions relative to the rotating frame, it is convenient to regard their resultant as a single force g, called the "apparent" gravitational force; i.e., $g = g^* + \Omega^2 R$.

For purposes of numerical calculation, the above vector equation must be decomposed into three scalar equations, corresponding to its components in three different coordinate directions. For ease of application to meteoro-logical data, they are usually chosen to lie parallel to the hypothetical sea surface and pointing northward, parallel to the surface and pointing east-ward, and perpendicular to the surface and pointing upward.

Thus, calculating the components in these three directions and bearing in mind that g is directed normal to the surface, we find that

$$\frac{du}{dt} - \frac{uv \tan \phi}{a} + \frac{uw}{a} - 2\Omega v \sin \phi + 2\Omega w \cos \phi = P_x + F_x$$

$$\frac{dv}{dt} + \frac{u^2 \tan \phi}{a} + \frac{vw}{a} + 2\Omega u \sin \phi = P_y + F_y$$

$$\frac{dw}{dt} - \frac{u^2 + v^2}{a} - 2\Omega u \cos \phi = P_z + F_z - g$$

Here x, y, and z are distances measured toward the east, north, and upward, respectively; u, v, and w are the corresponding components of U', the velocity relative to the rotating frame; ϕ is approximately the geographical latitude; and a is very nearly the mean radius of the earth. P_x, P_y, and P_z are the components of the pressure-gradient force (per unit mass) in the x, y, and z directions, respectively; F_x, F_y, and F_z are the corresponding components of the viscous force.

Applying the method of scale analysis to the above equations, we compare the magnitudes of the various types of terms. We first note that expansion of the total derivatives of u and v as sums of their partial varia-tions gives rise to terms whose dimensions and magnitudes are those of $2V^2/L$, where V is a characteristic horizontal wind speed and L is a char-acteristic distance between successive maxima and minima of the velocity field. Such terms are to be compared with $uv (\tan \phi)/a$ and $u^2 (\tan \phi)/a$, whose magnitudes are those of V^2/a. Thus, since the characteristic half wavelength of the large-scale weather disturbances is of the order of 1000 miles and the radius of the earth is about 4000 miles, terms of the size of $2V^2/L$ are an order of magnitude larger than those whose size is V^2/a.

Terms of the order of $2V^2/L$ also are to be compared with uw/a and vw/a, whose magnitudes are those of VW/a, where W is a characteristic vertical air speed. Now the characteristic horizontal wind speed lies in the range 10^3 to 10^4 cm/sec, whereas the characteristic magnitude of the vertical air speed associated with the large-scale disturbances lies in the range of 1 to 10 cm/sec. Thus terms of the size of $2V^2/L$ are about four orders of magnitude greater than those whose size is VW/a.

A good approximation to the horizontal equations may be had by omitting those terms that contain products of velocity components, so that those equations take the form

$$\frac{du}{dt} - fv = P_x + F_x$$

$$\frac{dv}{dt} + fu = P_y + F_y$$

where f, the so-called "Coriolis parameter," is $2\Omega \sin \phi$. It should be noted that the terms omitted are exactly those which arose from spatial variations in the directions of the coordinate axes; hence the equations are of the same form as the equations of motion in a cartesian system of coordinates.

Following a similar line of reasoning for the vertical motions, we can also estimate the magnitudes of the terms on the left-hand side of the dw/dt equation. Their magnitudes are, respectively, those of $2VW/L$, V^2/a, and ΩV. For the *large-scale motions* of the atmosphere, these quantities are of the order of 10^{-4}, 10^{-2}, and 1 cm/sec^2. The gravitational acceleration g, on the other hand, is of the order of 10^3 cm/sec^2. Accordingly, in applying the equation to the large-scale motions of the atmosphere, the entire left-hand side may be omitted. In fact, since the vertical accelerations due to molecular viscosity are also very much less than the gravitational acceleration, the dw/dt equation reduces to a statement that the gravitational acceleration g is almost exactly balanced by the acceleration P_z due to the atmosphere's buoyancy, i.e., $P_z - g = 0$. In other words, for meteorological purposes it is permissible to treat the atmosphere as if it were always in a state of hydrostatic equilibrium.

Owing to the large horizontal scale of weather disturbances, it turns out that the horizontal viscosity (due to the molecular transfer of momentum) are also negligible when compared to the Coriolis and horizontal pressure-gradient forces.

Expressing the pressure-gradient force in terms of the pressure field, we get the approximate form of the equations of motion usually employed in the hydrostatic approximation:

$$\frac{du}{dt} - fv + \frac{1}{\varrho} \frac{\partial p}{\partial x} = 0$$

$$\frac{dv}{dt} + fu + \frac{1}{\varrho} \frac{\partial p}{\partial y} = 0$$

$$\frac{\partial p}{\partial z} + g\varrho = 0$$

Since the variations in the directions of the coordinate axes are negligible, it is possible to combine the first two equations into a simple two-dimensional vector equation:

$$\frac{dV}{dt} + k \times fV + \frac{1}{\varrho} \nabla p = 0$$

where V is the projection of the velocity vector on a *horizontal* surface, k is a unit vector directed vertically upward, and ∇ is the *horizontal* vector gradient.

Applying scale analysis to this equation, we estimate the magnitudes of the first and second terms to be of the order of $2V^2/L$ and ΩV, respectively. If V is taken to be 10^3 cm/sec, the first of these is about 10^{-2} cm/sec^2, whereas the latter is about 10^{-1} cm/sec^2. Although such crude estimates certainly do not show that the horizontal accelerations of air are negligible, they do demonstrate that there is a tendency for the Coriolis and horizontal pressure-gradient forces to balance each other. Approximately, then, $V = k \times (1/\varrho f) \nabla p$. This is the so-called "geostrophic-wind" relationship. It shows that the winds tend to blow parallel to the isobars (lines of constant pressure) in surfaces of constant height and that their speed is roughly proportional to the horizontal gradient of pressure.

3. The Equations of Continuity

A principle essential to the theory of large-scale atmospheric motions is the law of mass conservation. It is expressed in mathematical terms most simply by considering the mass budget within a very small rectangular parallelepiped, or box, whose edges lie parallel to the coordinate axes and whose dimensions are Δx, Δy, and Δz in the x, y, and z directions, respectively. Now, since mass is neither created nor destroyed (in the absence of violent nuclear reactions), the net rate of mass transport into the box must be just the rate at which mass is accumulating in the box. That is,

$$\frac{\partial \varrho}{\partial t} = -\frac{\partial}{\partial x}(\varrho u) - \frac{\partial}{\partial y}(\varrho v) - \frac{\partial}{\partial z}(\partial w)$$

Another more useful form of the continuity equation may be derived by performing the differentiations indicated symbolically on the right-hand side of the equation. Rearranging terms, we find that

$$\frac{\partial \varrho}{\partial t} + u \frac{\partial \varrho}{\partial x} + v \frac{v \varrho}{\partial y} + w \frac{\partial \varrho}{\partial z} + \varrho \left(\frac{\partial u}{\partial x} + \frac{\partial v}{\partial y} + \frac{\partial w}{\partial z} \right) = 0$$

or, since the sum of the first four terms above is simply the total derivative of ϱ,

$$\frac{d\varrho}{dt} + \varrho\left(\frac{\partial u}{\partial x} + \frac{\partial v}{\partial y} + \frac{\partial w}{\partial z}\right) = 0 \quad \text{or} \quad \frac{d\varrho}{dt} + \varrho \operatorname{div} U = 0$$

If we were dealing with an *incompressible* fluid, the continuity equation would reduce to

$$\frac{\partial u}{\partial x} + \frac{\partial v}{\partial y} + \frac{\partial w}{\partial z} = 0 \quad \text{or} \quad \operatorname{div} U = 0$$

The momentum and mass conservation equations above contain horizontal derivatives that have been left in terms of local x, y, z coordinates. The modifications required to make the equations applicable to a global (or hemispheric) calculation is discussed later as the "mapping problem." In general, one either expresses the vector quantities directly in spherical coordinates before creating the numerical difference equations, or the x, y variables are kept for numerical differencing but replaced by mx, my, where m gives the mapping scale factor resulting from projecting the flat xy plane to a spherical surface.

4. The Equation of Energy Conservation

A physical principle essential to a theory of dynamical prediction is the first law of thermodynamics. If for large-scale motions the atmosphere can be treated as a nonviscous fluid, the only work it can do is to expand against normal or pressure forces. Thus the amount of heat energy added to a given mass of air must be exactly balanced by the change in its internal energy plus the work done in increasing its volume. This principle, the first law of thermodynamics, is stated mathematically in the following equation:

$$\frac{dq}{dt} = C_v \frac{dT}{dt} + \frac{dV}{dt}$$

Here dq/dt is the rate at which heat is added to a unit mass of air (dq is written with a mark through the d as a reminder that it is not necessarily an exact differential) and C_v is the specific heat of air for constant volume. Since the internal energy of "perfect" gases depends only on absolute temperature T, the first term on the right-hand side of the equation is the rate at which the internal energy of a unit mass of air increases. The second term, in which V is the volume per unit mass, is the rate at which the unit mass does work in expanding against pressure forces.

A special case of particular interest to meteorology is the adiabatic process, i.e., a process by which the temperature, pressure, and specific volume may change, but without the addition of heat energy. In this case

$$C_v \frac{dT}{dt} + p \frac{dV}{dt} = 0$$

This equation can be integrated by making use of the ideal gas law equation of state $pV = RT$, where R is the gas constant for air (here assumed to be unsaturated with water vapor). Differentiating the equation above and substituting in for $p\,dV/dt$, we find that

$$C_p \frac{dT}{dt} - V \frac{dp}{dt} = 0$$

where C_p, the specific heat for constant pressure, is equal to $(C_v + R)$.

If we let p_0 equal 10^3 millibars (10^6 dynes/cm²), the quantity $T(p_0/p)^k$ is known as the "potential temperature" and is generally designated by the symbol θ. It is simply the temperature that an individual element of air would take on if it were lowered or raised adiabatically to a pressure of 1000 millibars, starting with its current pressure p and absolute temperature T. The first law of thermodynamics for adiabatic processes now can be stated very concisely by writing $d\theta/dt = 0$. According to this equation, if a process is adiabatic the potential temperature of any individual element of air will remain unchanged.

In the general case of nonadiabatic flow, the individual changes of potential temperature are given by

$$\frac{d\theta}{dt} = \frac{\theta}{C_p T} \frac{dq}{dt}$$

Except in precipitating clouds of limited vertical or horizontal extent, in relatively thin layers of heat-absorbing or radiating cloud, and in shallow layers of air next to a relatively warmer land or sea surface, the individual changes of *potential* temperature are no more than a few degrees centigrade per day. On the other hand, owing to the large vertical displacements of air in the free atmosphere, the changes of *absolute* temperature experienced by individual fluid elements are of the order of 20°C per day. For some purposes, such as forecasts of one to three days, it is permissible to treat the large-scale flow of the atmosphere as if it were adiabatic, but not for long periods of time.

5. Some Nonadiabatic Effects—Friction and Precipitation

In the formulation of the hydrostatic equations of motion and the geostrophic approximation, the effects of friction were barely mentioned. Similarly, the heating term q in the energy equation was included with no discussion as to its significance. These, along with the equation of state changes due to precipitation, are the nonadiabatic effects that "spoil" the mathematics of dynamic models.

Three processes are effective in heating the atmosphere: (1) radiative transfer, (2) release of latent heat caused by large-scale organized patterns of rising motion, and (3) turbulent transport of sensible heat and release of latent heat by small eddies. A satisfactory numerical treatment is still being sought for all three processes.

a. The Importance of Nonadiabatic Effects. As Phillips[5] points out, it would seem at first glance to be a serious mistake to do any calculations with the adiabatic equations, since the unequal heating of the atmosphere is the ultimate driving agency for all motions of the atmosphere (except for very small tidal effects). However, many estimates of both the input and dissipation of energy in the atmosphere have shown that these are for the most part very slow-acting forces and that it would take one or two weeks for the existing kinetic energy of the atmosphere to decrease by a factor of e^{-1} even if all heating were stopped. For forecast periods of two or three days, it is possible to neglect q and F entirely.

The main exception to this rule is the heating by the release of latent heat in precipitation. This process frequently seems to be important in affecting the motion over a limited area even for one- or two-day forecasts and, of course, the prediction of rainfall is one of the most important goals of weather forecasting.

For longer range forecasts, on the other hand, both q and F are important. Unfortunately, much of the large-scale heating of the atmosphere, as well as the effect of friction on the large-scale motions, occurs through the cumulative effect of motions on a much smaller space and time scale, e.g., cumulus clouds. Since it is impractical to forecast the details of these small turbulence elements, their effect on the large-scale motions must be approximated by some type of turbulence or boundary layer theory.

b. Friction. The main effect of friction in the atmosphere is concentrated in the layer of air next to the ground. The lower atmosphere can be thought of as consisting of three layers: the surface boundary layer (Prandtl layer), about 50 to 100 meters thick; the planetary boundary layer (Ekman layer), which extends to 500 to 1000 meters; and (3) the free atmosphere above this. The free atmosphere also consists of layers: the troposphere, up to about 10 km; the stratosphere, which extends to about 100 km; and the ionosphere, above 100 km. A very simple representation of this effect can be obtained if a term $F' = \partial\tau/\partial z$ is retained in the horizontal equations of motion. τ represents the horizontal stress exerted on a layer of air by the layer of air immediately under it and is to be thought of as representing the effect of small eddies in redistributing momentum. When vertical differences are then introduced into forecast equations, the effect of surface friction can be allowed for by simply setting τ equal to zero everywhere except at the ground, where a relation such as $\tau = -kV$, or $\tau = -k' \mid V \mid V$, can be applied. An alternative derivation is to use the Ekman theory to estimate the frictionally induced flow across the isobars in the surface friction layer ($\sim$1 km). This drift is from high to low pressure, giving horizontal convergence in regions of low pressure and divergence in high pressure centers. The resulting vertical velocities at the top of the friction layer can then be incorporated in the lower boundary condition.

c. Precipitation. It is possible to compute the release of latent heat by precipitation in one- or two-day forecasts, when the addition of new water vapor to the atmosphere may be either crudely approximated or neglected. These calculations have been based on the assumption of a pseudo-adiabatic condensation process:

$$\frac{q}{T} = \frac{ds}{dt} = \frac{dp}{dt}\frac{ds}{dp} = \omega\frac{ds}{dp}$$

In this process it is assumed that any condensed water vapor immediately falls out as precipitation. The coefficient ds/dp is then negative for expanding air ($\omega < 0$) but is equal to zero for either unsaturated air or for the case of saturated air with $\omega > 0$. An additional equation describing the transport of water vapor similar to the following must be used:

$$\frac{\partial\mu}{\partial t} = -V\cdot\nabla_p r - \omega\frac{\partial r}{\partial p} + \dot{r}$$

Here μ represents the specific humidity or the percentage by weight of water vapor in a volume of moist air. The negative of $\dot{r}$ is the rate of condensation and is therefore proportional to q, the rate at which latent heat is given up to the air. The test for the existence, or nonexistence, of saturation at a point is easily made by comparing r with the saturation mixing ratio r_s—a known function of T and p.

Closely associated with precipitation are clouds, which are also strongly influenced by radiation and vice versa.

d. Atmospheric Heating and Cooling by Radiation. The effects of radiative cooling in the atmosphere are less well formulated in existing models, but progress is being made.[6] Although the hydrodynamic energy of the atmosphere is dominant and its motion can be calculated adiabatically to a fairly reasonable approximation, one must keep in mind that the earth receives virtually all of its energy from space in the form of electromagnetic radiation from the sun. In turn, the earth radiates back into space almost exactly the same amount of energy, but at a lower temperature.

The transformation of the incident solar radiation into scattered and thermal radiation and the consequent thermodynamic effects on the earth's gaseous envelope, are very complicated phenomena, requiring the most advanced methods of molecular physics and quantum mechanical calculations. Absorption along a real atmospheric path, where pressure, temperature, and composition all vary, presents problems of which only a few have been solved (see Goody[7]). Fortunately theory exists[37] that permits the use of simpler approximations in weather models.

e. Calculation of Cloud Effects. If a cloud layer is present, the drops of water of which a cloud consists are comparable in size to the wavelengths of thermal radiation, and their number per unit volume is quite high. In this case consideration of the scattering is, therefore, of great importance. The accurate solution of the problem of radiative heat transfer in clouds can be obtained only by using a detailed equation for radiative energy transfer. Calculations indicate that a cloud is "active" with respect to thermal radiation only around its edges. The flux of thermal radiation entering the cloud is completely absorbed in a distance of a few tens of meters.

Probably the most formidable computational problems meteorologists are likely to face will arise from the calculation of cumulus convection. In such problems no simple hydrostatic approximation for vertical motion can be assumed. The phase changes of water in air will have to be carried in four forms—vapor, ice, water, and droplets. Even electrostatic forces between droplets may have to be calculated, at least in some average sense. These microscale calculations are as demanding of computer time as the general circulation problems.

f. Air-Sea Interface. Although the most important driving force for the atmosphere is ultimately the radiation from the sun, the second most important is no doubt the atmosphere's interactions with the ocean. The tremendously large heat capacity represented by the oceans of the earth provide its stability and relative uniformity of temperature. The oceans and the air can be considered a two-fluid system coupled relatively loosely but in a very important way.

A government report edited by Benton[8] on the interaction between the atmosphere and the oceans emphasized the fact that the atmosphere and the oceans together form a single mechanical and thermodynamic system and that an understanding of the way in which energy, gases, particles, and electric charges move across the interface between the two is essential to the development of geophysics. Energy transfer, in the form of radiation or latent heat, affects the circulation of both the atmosphere and the oceans.

As in the case of radiative cooling mentioned earlier, the air–sea interaction can be neglected for short-range forecasts. There are, however, cases in which the effect can be very important. Scientists at the Fleet Numerical Weather Center at Monterey, California, have studied these effects in a number of calculations. Laevastu[9] stresses the point that the development of truly synoptic oceanographic analysis and forecasting emphasizes thermal structure in the sea and requires the quantitative knowledge of energy exchanged between the sea and the atmosphere. Temperature in oceanographic analyses has the same importance as atmospheric pressure in meteorological analyses.

In general, regions of greatest evaporation are those wherein the northerly transport of surface water is greatest and which are subjected during winter to frequent invasions by cold, dry air masses from the interiors of continents. In the Pacific, this region is situated off Japan where cold, dry air of continental Asiatic origin frequently traverses the northward-flowing, warm current.

Figure 1 summarizes the major physical phenomena studied by a general circulation model with an estimate concerning how well each is understood.

g. Weather and Climate Modification. Certainly no discussion of the future of meteorological computations would be complete without mentioning the enormous implications of weather and climate modification. Several studies have been made in recent years concerning the general problems of weather modification. A National Academy of Sciences study headed by MacDonald[5] resulted in an excellent report on the problems and prospects of weather modification and is recommended to anyone interested in the subject.

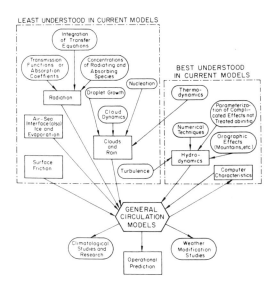

Fig. 1. Problem areas associated with general circulation models.

The major portion of atmospheric energy exchange is due to the release of instabilities inherent in the preferred states of the atmosphere. These dynamically unstable situations are looked upon as "levers" or "soft spots" in the system where man's efforts might be able to trigger a chain of nature reactions.

Man has exploited these instabilities on a limited scale in the belief that the effects would be as short-lived as the phenomena themselves and that the energies released would not escalate to the level that would change the weather permanently.

It is obvious from geological evidence that the earth–atmosphere system can support radically different climatic regimes, some of which could be disastrous to civilization. We do not yet know what can cause a shift from one climatic regime to another or whether change can occur in an "instant" of geologic time or only as a secular cyclic process; our few theories still hang on the most tenuous evidence.

Numerically integrated mathematical models of the atmosphere have come to be regarded as necessary tools for research in modification of the atmosphere. This is particularly true in areas where actual experimentation would be too costly, take too long, or possibly be irreversible.

II. NUMERICAL METHODS

A. Physical Considerations

Because the partial differential equations (or in the case of radiation, integral equations) for a fluid are nonlinear and possess difficult initial and boundary conditions, they must be solved by numerical methods for practical cases. This involves converting the equations describing the particular physical model under consideration into a form that can be solved by numerical algorithms on a computer.

Although it is a simple matter to convert a given system of differential equations to some finite difference form, it is a far more difficult task to obtain physically meaningful results even if the integration method proves to be stable. The hydrodynamical equations of motion contain among their solutions the high-speed sound and gravity waves which, meteorologically speaking, are spurious information. Unless an unrealistically small time step (several seconds) is used, these solutions have a tendency to amplify in time and overshadow the physically meaningful results. It is necessary to find a set of difference equations that will be stable in the sense that the calculation can go on for an indefinite time without nonsensical results developing. This is a particularly sensitive matter in an atmospheric model since there are conditions under which the atmosphere itself can be temporarily unstable so that small disturbances really do grow. This must be permitted in the numerical model also, but in a reasonable way.

By far the most serious obstacle to solving the hydrodynamic equations arises from the properties of the atmosphere itself. The large-scale horizontal accelerations of air are about an order of magnitude less than either of the forces per unit mass taken individually, i.e., the Coriolis force due to the rotating earth and horizontal pressure-gradient force. The atmosphere maintains itself near some state of balance and much more nearly so than is revealed by direct measurement.[10]

For similar reasons, the horizontal divergence of the wind velocity cannot be computed accurately from direct measurements of the wind. It can be inferred indirectly that the sum $(\partial u/\partial x + \partial v/\partial y)$ is generally an order of magnitude less than either $\partial u/\partial x$ or $\partial v/\partial y$ taken individually, i.e., the latter tend to compensate each other almost completely. Thus in order to compute $\nabla \cdot V = \partial u/\partial x + \partial v/\partial y$ to within 10% accuracy, the wind components must be measured to within 1% accuracy. Winds, however, are not measured and reported to within better than 10% accuracy. As a result the vertical air speed cannot be computed accurately unless spuriously large fluctuations of divergence can somehow be suppressed.

Present global general circulation models use equations in which the hydrostatic assumption is made. (This approximate set of equations is somewhat misleadingly referred to as the "primitive equations" in meteorological literature.) This is an accurate approximation for motions with horizontal scales of 25 miles or longer. Vertically propagating sound waves are excluded by this technique but gravity waves are retained. The size of the time step permitted is proportional to the horizontal space increment; for a latter value of 125 miles, the time step should be less than 10 minutes.

B. Finite Difference Solutions and Nonlinear Instability

1. What Is an Acceptable Solution?

At the present time no single finite-difference analog of the primitive equations has emerged that meets all the objectives of the meteorological community. Such a set of difference equations must be, in addition to being "physically" acceptable, mathematically accurate and stable. By accuracy is meant consistency, and by stability, convergence.

In 1956 Phillips[11] encountered an unexpected difficulty in his early attempt at long-term integration of the meteorological equations. After about 20 simulated days the solution began to show a structure termed "noodling," in which the motion degenerated into eddies of elongated, filamented shapes. Once formed, the eddies intensify without limit, causing a nonlinear computational instability and explosive growth of the total kinetic energy. Phillips showed that the instability is caused by "aliasing" or misrepresentation of the shorter waves because a finite grid cannot properly resolve them. Phillips showed further that the instability could not be reduced by shortening the time interval.

2. Nonlinear Instability in the Leapfrog Differencing Scheme

Using the equation $\partial u/\partial t = u(\partial u/\partial x)$ as an example, the so-called "leapfrog" difference equation is (see Gary[38])

$$u_j^{(n+1)} = u_j^{(n-1)} + \frac{\Delta t}{\Delta x} u_j^{(n)}(u_j^{(n)} - u_{j-1}^{(n)})$$

where $u_j^{(n)} = u(x_j, t_n)$ with $x_j = j\Delta x$, $t_n = n \Delta t$, and $\Delta x = 1/M$.

A difference scheme is stable if there is a constant K such that $\| u^{(n)} \| < K \| u^{(0)} \|$ for all n, where $\| u \|$ denotes some suitable norm. This inequality must hold for all sufficiently small Δx. Usually it is necessary to add the requirement that $\Delta t \leq c \Delta x$ for some constant c in order to obtain stability.

If the scheme above is linearized (formally replacing $u_j^{(n)}$ by $c + u_j^{(n)}$, where c is constant), we obtain

$$u_j^{(n+1)} = u_j^{(n-1)} + \frac{\Delta t}{\Delta x} c(u_{j+1}^{(n)} - u_{j-1}^{(n)})$$

If the boundary conditions are ignored it can be shown that this scheme is stable provided $\Delta t |c| / \Delta x \leq 1$. Yet certain solutions of the nonlinear leapfrog equations become unbounded as n increases regardless of the value of Δt. If Δt is small enough, the linearized equations will be stable, but the nonlinear equations will be unstable. There are, however, techniques to eliminate this nonlinear instability.

One general principle that can be applied to the construction of these difference schemes is the requirement that the difference scheme conserve certain integral invariants of the differential equation. For example, the differential equation $\partial u / \partial t = u(\partial u / \partial x)$ implies that the moments

$$\int_0^1 [u(x, t)]^k \, dx = M_k(t)$$

are constant for $k = 1, 2, \ldots$

Suppose we consider the leapfrog finite-difference scheme

$$u_j^{(n+1)} = u_j^{(n-1)} + \frac{\Delta t}{\Delta x} u_j{}^n (u_{j+1}^n - u_{j-1}^n)$$

If we sum both sides of this equation, cancellation on the right-hand side produces the identity

$$\sum_{j=0}^{M} u_j^{(n+1)} = \sum_{j=0}^{M} u_j^{(n-1)}$$

Note that certain types of boundary conditions will not allow this relation, i.e., there may be a net flux through the boundary. With periodic boundary conditions the finite difference scheme conserves the first moment, provided

$$\sum_0^{M} u_j^{(1)} = \sum_0^{M} u_j^{(0)}$$

3. Nonlinear Instability in the Implicit Difference Scheme

Next we consider the implicit difference scheme

$$u_j^{(n+1)} = u_j^{(n)} + \frac{\Delta t}{2\Delta x} u_{j+1}^{(n+\frac{1}{2})} [u_{j+1}^{(n+\frac{1}{2})} - u_{j-1}^{(n+\frac{1}{2})}]$$

where $u_j^{(n+\frac{1}{2})}$ denotes $(u_j^{(n+1)} + u_j^{(n)})/2$. Note that $n + 1$ quantities appear on both sides of the equation. The solution of this nonlinear equation for $u_j^{(n+1)}$ can cause difficulty.

The first moment is conserved for this difference scheme, i.e.,

$$\sum_{j=0}^{M} u_j^{(n+1)} = \sum_{j=0}^{M} u_j^{(n)}$$

However, the second moment, $\sum_{j=0}^{M} [u_j^{(n)}]^2$, is not constant.

The above difference scheme is unconditionally stable (in the L_2 norm). By unconditional stability we mean that it is stable for all values of Δt. The Courant–Friedrichs–Levy argument shows that an explicit scheme for a hyperbolic system cannot be stable for all values of Δt (with Δx fixed). The algebraic argument is used to show that the second moment preserved is usually indifferent to the value of Δt. Therefore it is unlikely that an explicit difference scheme will preserve the second moment. However, if there is more than one dependent variable in a hyperbolic system of equations, then it may be possible to preserve the second moment of some, but probably not all, of these dependent variables. The boundary values also become more difficult.

4. Arakawa's Method for Nonlinear Stability

An approach explored by Arakawa[3,12] attacks the nonlinear instability problem at its source. The method is based on a quasi-conservation of these moments. The finite-difference approximations used at various grid points are interrelated, so that what disappears from one grid point reappears somewhere else. This provides a discrete analog of the integral constraints governing mean kinetic energy, mean square potential temperature, etc. It does not prevent aliasing, but the aliasing errors do not amplify. The following is a description of Arakawa's method as it applies to our simple equation.

Suppose we difference the equation $\partial u/\partial t = u(\partial u/\partial x)$ in space but not in time. That is, we consider a system of equations

$$\frac{\partial u}{\partial t} = \frac{1}{3\Delta x} [u_j(u_{j+1} - u_{j-1}) + (u_{j+1})^2 - (u_{j-1})^2]$$

where $u_j = u_j(t) = u(x_j, t)$ with $x_j = j\Delta x$, $0 \leq j \leq M$, $u_0(t) = u_M(t)$. It can be shown that the first and second moments

$$\sum_{j=0}^{M} u_j(t) \quad \text{and} \quad \sum_{j=0}^{M} [u_j(t)]^2$$

are conserved. Both Arakawa[12] and Lilly[13] provide methods for deriving such conservative difference schemes. Fromm[40] also has done much original work in this field. (See the references in his chapter in this book.)

The equation still must be differenced in time. If an explicit time difference such as the leapfrog scheme is used, the second moment will not be conserved exactly. However, the scheme is not nearly as prone to nonlinear instabilities as the straightforward leapfrog scheme. Experiments show it still may be necessary to do some time averaging in order to avoid nonlinear instabilities. That is, every 50 (approximately) time steps the values of $u_j^{(n+1)}$ may be replaced by $\frac{1}{3}(u_j^{(n+1)} + u_j^{(n-1)})$. This seems to remove a slight tendency toward nonlinear instability.

5. The Lax-Wendroff Difference Scheme

The Lax–Wendroff[14] method has become well known as a very stable method of computing solutions to partial differential equations. Using the equation $\partial u / \partial t = 1/2(\partial^2 u)/(\partial x^2)$ as an example, the first step of the method is to compute

$$u_{j+\frac{1}{2}}^* = \frac{1}{2}(u_j^n + u_{j+1}^n) + \frac{\Delta t}{2\Delta x}[(u_{j+1}^{(n)})^2 - (u_n^{(n)})^2]$$

and then to define $u_j^{(n+1)}$ by

$$u_j^{(n+1)} = u_j^{(n)} + \frac{1}{4}\frac{\Delta t}{\Delta x}\left[(u_{j-\frac{1}{2}}^*)^2 + \frac{(u_{j+1}^{(n)})^2 - (u_{j-1}^{(n)})^2}{2}\right]$$

This scheme is apparently free from nonlinear instability although nonlinear instability has been produced under circumstances that probably would not arise in normal use of the method. Note that the scheme has second-order accuracy in both space and time.

This scheme is dissipative, which probably accounts for the freedom from nonlinear instability. If the differential equations comprising the model are slightly unstable for certain wavelengths (which is true for the weather) and a dissipative finite difference scheme is used that does not resolve these wavelengths very accurately, while dissipating them, then the finite difference model may not show the true instability in the differential equations. In other words, the model will not produce much real weather. In real fore-casts the use of a dissipative scheme may be unwise, although if the mesh were fine enough then the dissipative scheme should work satisfactorily. Kasahara and Washington use a neutral stability leapfrog method, with one Lax–Wendroff[11] step every 45 time steps. This is sufficient to stabilize the calculation. Both Smagorinsky and Leith avoid nonlinear instability by putting in damping coefficients that diffuse away singularities.

6. The Leith Fractional Time Step Method

We will now describe a difference scheme that has been used by Leith[15] in his general circulation computation. The essence of this scheme is that it enables a stable difference scheme for a one-space-dimensional problem to be generalized to multidimensional problems. For example, suppose we have a difference scheme for the equation $(\partial u/\partial t) = u(\partial u/\partial x)$ that related $u^{(n+1)}$ and $u^{(n)}$. We might write this scheme in the form $u^{(n+1)} = L_x(u^n)$, where L_x is some nonlinear operator in the vector $[u_j^{(n)}]$. For example,

$$u_j^{(n+1)} = L_x(u_j^n) = u_j^n + \frac{\Delta t}{\Delta x} u_j^n(u_j^n - u_{j-1}^n)$$

is an operator of first-order accuracy.

The method of fractional time steps constructs a difference approxima-tion as follows:

$$u^{(n+\frac{1}{2})} = L_x(u^{(n)})$$

$$u^{(n+1)} = L_y(u^{(n+\frac{1}{2})}) = L_y[L_x(u^{(n)})]$$

If the operators L_x and L_y are stable (i.e., $\| L_x \| \leq 1$ and $\| L_y \| \leq 1$), then the composite operator will be stable. This fact is the beauty of the method. Also, if L_x and L_y conserve a certain moment then the composite operator also will conserve it. However, even though each of the operators L_x and L_y has second-order accuracy in time [the truncation error is $0\ (\Delta t^3)$], the composite operator may be only first order.

C. Existing General Circulation Models

At the present time there are several research groups in the United States actively working on general circulation models.[6,19,20] Abroad there is one large group working in the Soviet Union at Novosibirsk. Many other groups here and abroad are working on the theoretical aspects of general circulation or are using numerical flow calculations for operational forecasts. Research in general circulation calculations obviously is heavily dependent upon the availability of large computers and upon the existence of a wealthy sponsor (usually some government agency).

The present researchers in the field all owe a great debt to the work of Charney and Phillips[16] at the Institute for Advanced Study in the early 1950's. They also owe much to the early operational models of Shuman[17] and Cressman and Bedient,[18] who verified the importance of numerical prediction. Operational numerical weather prediction first came into its own with the formation of the Joint Numerical Weather Prediction unit (JNWP) in 1954. The U.S. Weather Bureau, the Air Force, and the Navy jointly established the JNWP to capitalize on the research[10] that had been done at the Institute for Advanced Study and the Air Force's Cambridge Research Center and under the late C. G. Rossby at Stockholm. The three main operational groups in existence today in the United States are direct descendants of the JNWP. They are The ESSA Weather Bureau's National Meteorological Center at Suitland, the 3rd Weather Wing at Offutt AFB, and the Fleet Numerical Weather Center at Monterey.

Of course, there is a tremendous difference between doing research on general circulation and issuing operational forecasts every 12 hours. The pressure of the latter has forced the use of simpler models that have been tried and found reliable under operational conditions. The operational models are steadily expanding as techniques are improved by research and as better data become available. For more details see, e.g., the papers by Cressman,[10] Wolff,[21] O'Neill,[22] Stauffer,[23] and their collaborators.

The main difference between research and operations can be related to the "real data" problem. The operational groups are geared to the world-wide data acquisition and communication networks that furnish the initial conditions for their prediction models. A very important phase of data collection is the verification and smoothing of the raw measurements for the entry into the numerical model. For example, the National Meteorological Center uses a combination of computers to perform this task. The distribution of the completed products of the centers to their customers is another large topic we can only mention here.

D. Nonfinite Difference Methods in Numerical Weather Calculations

In investigating the future computer requirements of numerical weather calculations, one always has the uneasy feeling that some radical departure from the present finite-difference methods will make the computing estimates completely invalid. The hope is that some entirely different numerical method might reduce the computations required by many orders of magnitude.

One suggestion that arises repeatedly is that spectral methods, such as Fourier transform methods, might be employed for the calculations. This was considered as early as 1950 by Charney, Fjortoft, and von Neumann[21] as a means for solving the Poisson equation arising in their geostrophic model. The necessity of solving the Poisson equation on each time step, which arises in the geostrophic models, is being abandoned by the primitive equation approaches currently in vogue.

Fourier methods, however, have continued to be used to analyze and treat stability problems in numerical procedures.[25,26] These yield a good description of the nonlinear instability arising from a cascading of energy from low- to high-frequency waves due to the "aliasing" mentioned earlier. This phenomenon results from the fact that products of functions yield Fourier components outside the spatial frequency range corresponding to the grid size. These components then contribute erroneously to the frequencies, modulo the number of grid points, lying in the spectrum corresponding to the grid. This suggests filtering out these high frequencies by the introduction of diffusion terms in the equations or by periodically sweeping over the grid with some averaging process. But it also suggests a less artificial means, namely, the carrying out of the calculation in the spatial frequency domain with the simple expedient of dropping Fourier components outside the frequency range being considered.

The most serious objection to the use of spectral methods is that they do not easily accept modifications, such as the introduction of nonconstant coefficients or extra terms, which can be just "tacked onto" a finite-difference formulation. This property, plus the difficulty to read and understand the spectral equations, makes their acceptance somewhat doubtful in the near future. Nevertheless, their great potential payoff keeps interest alive.

III. COMPUTER REQUIREMENTS

A. High-Speed Computer Characteristics

Is is futile, of course, to try to describe a computer in terms of a single speed number or to quote storage in terms of a single value.[27] However, if such numbers are not taken too seriously, they can be helpful in giving an idea of the capabilities of these devices. The secret is to emphasize the

important effects, i.e., the ones whose inclusion results in a factor of two
or more, while trying not to get lost in the myriad 10% effects.

Figures 2 and 3 show the progress of large machines versus time, both
in the raw speed (measured in terms of millions of instructions per second
executed) and in the internal storage (rated in terms of characters of storage
regardless of size of the words). Word size itself can affect useful storage
by as much as a factor of two but usually it does not have this effect for
scientific applications. The graphs show a steady increase in speed and
storage and, if one overlooks occasional plateaus and fluctuations, there is
almost a constant rate of increase of performance and storage. Experts in
the computer field have been predicting a coming saturation in computer
speed and size ever since the beginning. It has not yet happened, however,
and there are no indications that it will necessarily happen in the next
generation of computers.[28–30,41] The velocity of light, which is often quoted
as the main barrier to computer speed, is still there, but it has not proved
to be the barrier we expected. There are ways of obtaining an effectively
high instruction operation rate other than simply making the individual
components go faster and faster.

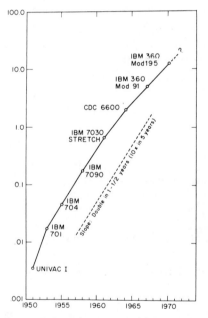

Fig. 2. The growth of computer speed versus
time (computer speed in millions of average
instructions per second, MIPS).

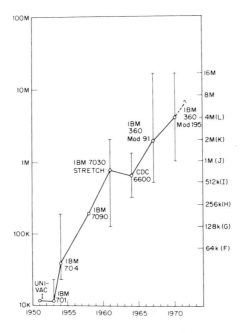

Fig. 3. The growth of computer storage versus
time (primary computer storage size in characters showing range available on each machine
and the usual value).

Since the beginning of the development of modern computers there
has always been at least one machine at any given time that could be called
the "supercomputer" of its period. These machines are always "stretching
the state of the art" of the technology at their time. The STRETCH computer (IBM 7030) obtained its name from this phenomenon. Every supercomputer built to date, and probably most new ones to be discussed in the
future, has one thing in common, i.e., their protagonists use the numerical
weather problem as one of the reasons for developing their machines. It has
the following advantages for this purpose: It can saturate the biggest computer. It is very important for the nation to solve it. It resembles many other
problems based on partial differential equations and is completely un-
classified from the national security point of view.

A cynic might say that numerical weather prediction seems to be a
"feedback loop" in which the careful analysis of physical theories and
numerical methods is aimed mainly at justifying the calculations the available machine can do.

B. Need for Supercomputers

It was pointed out in a National Academy of Sciences report[5] that there are a number of problems within the context of geophysical hydrodynamics that are now being worked on that may provide us with the basis for an estimate of the computation demands that can be anticipated within the next decade. The characteristic time and space scales range from those concerning turbulent exchange processes to those responsible for maintaining the large-scale ocean circulations. See Table II.

It is of interest that, despite a span of simulated-experiment time ranging from 10 sec to 30 years (a ratio of 10^{-8}), the number of dependent variables generated during the experiment is roughly invariant, i.e., 10^{10}. Thus the problems are of similar computational magnitude. It is also typical of and common to such hydrodynamical calculations that approximately 200 computer operations are required to generate one dependent variable at each new time step so that somewhat in excess of 10^{12} computer operations are necessary to complete a single experiment.

We must define a "reasonable time" to spend in doing a numerical simulation experiment. The purpose of doing experiments is to study the nonlinear response of the numerical model to changes in parameters. By this process one develops the sought-after insight. Since a series of such experiments is necessary in order to span a physically realizable range of each of the parameters, we can arrive subjectively at a threshold of tolerance. Obviously 1000 hours of machine time (one-half year of first-shift time) for a single experiment is intolerable. The threshold is more likely to be 100 hours (2.5 weeks of first-shift time), but a convenient time is closer to 10 hours. Operational considerations also impose similar limits. Hence we need to be able to perform 10^{12} computer operations in 10 hours, or approximately one operation every 30 nsec.

C. Estimates of Computer Requirements in Meteorological Calculations

1. Main Properties Contributing to Speed and Storage Requirements

A really accurate estimate of the computer speeds needed for future global weather calculations is probably not possible no matter how much detail one puts into it. Even though the estimated average values may be fairly accurate, the overall speed of a system often depends upon combina-

Table II. Space Resolution, Dimensionality and Pertinent Dependent Variables for Four Classes of Problems Requiring Very Large Computers[5]

	Type of simulation experiment			
	Turbulence	Convection	Atmospheric general circulation	Ocean general circulation
Filtering approximation	Boussinesq	Boussinesq	hydrostatic	geostrophic
Dimensionality	3	3	3	3
Dependent variables	$4\ (u, v, w, T)$	$5\ (u, v, w, T, r)$	$4\ (u, v, T, r)$	$4\ (u, v, T, S)$
Number vertical levels	(Estim. 30)	100	10	10
Number horizontal mesh points	(Estim. 100^2)	100^2	100^2	100^2
Time step[a]	(Estim. 0.001 sec)	5 sec	5 min	5 hr
Simulated experiment time	(Estim. 10 sec)	3 hr	100 days	30 years
Dependent variables/time step	10^6	5×10^3	4×10^5	4×10^5
Time steps/experiment	10^4	2×10^2	3×10^4	6×10^4
Dependent variables/experiment	10^{10}	10^{10}	10^{10}	2×10^{10}

[a] Dictated by the space resolution and the linear computation stability requirements.

tions of interactions. In other words, system performance depends on the distribution of the variables and their correlations as well as their mean values. An unfortunate coincidence of slow events can spoil the running time of the whole calculation, thus making accurate estimates very difficult. Even so, one can usually arrive at numbers bracketing the computer speeds and requirements that are reasonably safe as typical values.

For the global weather calculations we are considering, the following appear to be the main properties that contribute to the computer speed and storage requirements. Some trade-off between them is possible:

1. The complexity of the numerical model being used, including the number of physical effects being approximated and the difficulty of each algorithm used to compute them.

2. The number of computer operations needed per program and their types compared to those available on the computer under consideration.

3. The total number of space points and time steps needed in a calculation.

4. The desired ratio of speed of the computed solution of the atmospheric motion compared to real time.

There are other important factors one could list, such as ease of programming, reliability, convenience of the display of results, etc. These are very important in building a satisfactory system but have only a secondary effect on the estimates of raw computer speed and storage capacity.

2. Complexity of the Model

The most important property that gives the character of the calculation is, of course, the numerical model of the weather problem being solved. It is made up of mathematical approximations to the physical quantities, the particular numerical algorithms being used to solve the mathematical equations, and the interactions allowed between the equations being solved.

It is customary to consider the dependent variables in a problem as either prognostic or diagnostic variables. There are five fundamental prognostic variables that must be evaluated and stored each time for each mesh point in the three-dimensional grid: the three components of wind velocity, the temperature, and the water vapor content. In addition, there are two or more quantities that are two-dimensional in nature. These are kept in storage corresponding to the horizontal grid. An example would be the surface pressure.

In addition to the variables there are large numbers of constants. Usually no three-dimensional constants are carried, but there can be a number of two-dimensional arrays of constants. Surface properties such as the differences between land and sea, the number of daylight hours, radiation properties, etc. can be kept as constants.

Because of the number of special two-dimensional cases concerned with the surface conditions it is easier to describe the number of quantities for a vertical column of points rather than for each point in the three-dimensional mesh. On this basis Leith's program[15] stores 31 variables per horizontal point containing 6 vertical levels and Smagorinsky[6] uses 37 variables for 9 levels plus other diagnostic variables not computed every cycle. Table III contains a typical list of parameters and their dimensions.

3. Number of Computer Operations Executed

For rough estimates the best parameter seems to be the number of instructions executed per mesh point per cycle. This is a complicated average to compute from a program, but it is easy to measure and very easy to use in making estimates. After examining several programs and making allowances for cases of input output limitations, the following seem to be typical.

For a straightforward hydrodynamics program with no complications, approximately 300 instructions are executed per mesh point in a three-dimensional grid per time step. The most complicated programs with many complex physical effects included run less than 3000 instructions executed per mesh point per time step. The programs themselves could be different in total size by more like a factor of 100. The important point is that the running time of a calculation in terms of computer operations per point per cycle does not seem to range over more than a factor of 10 in practice.

4. Mapping Problem

One problem somewhat unique to the general circulation model is that of choosing a coordinate system to represent the surface of the earth in a convenient fashion in the computer. At first glance it seems that a trivial change of coordinates should solve it. However, like many such "trivial" matters it has caused a great deal of serious thought and planning, since the whole structure of the problem storage and the details of the difference equations depend on the form of mapping used.

Three types of mappings commonly are used. The easiest to understand is the latitude-longitude grid such as that used by Leith and Mintz. It has the difficulty that the longitude lines get closer together as one goes toward

Table III. List of Typical Parameters Used in Global Weather Calculations[a]

	Variable name	Dimension[b]

1. Prognostic variables

u v	horizontal velocity components	$N \times E \times V$
w	vertical velocity component	$N \times E \times V$
μ	water vapor variable	$N \times E \times V$
T	temperature	$N \times E \times V$

2. Diagnostic variables (usually not evaluated every cycle)

Φ	geopotential		$N \times E \times V$
F_x^h F_y^h	momentum due to horizontal diffusion		$N \times E \times V$ $N \times E \times V$
F_x^v F_y^v	momentum due to vertical diffusion		$N \times E \times V$ $N \times E \times V$
$\dot{g}_h$		horizontal diffusion	$N \times E \times V$
$\dot{g}_v$	rate of heat	vertical diffusion	$N \times E \times V$
$\dot{g}_r$	generation due to:	radiation	$N \times E \times V$
$\dot{g}_t$		thermodynamic transformations	$N \times E \times V$
$\dot{W}_h$		horizontal diffusion	$N \times E \times V$
$\dot{W}_t$	rate of water vapor	vertical diffusion	$N \times E \times V$
$\dot{W}_t$	generation due to:	thermodynamic transformations	$N \times E \times V$
CLA	cloud amounts		$N \times E \times 3$
CT	cloud tops		$N \times E \times 3$
CB	cloud bottoms		$N \times E \times 3$
CA	cloud absorptivities		$N \times E \times 3$
CR	cloud reflection coefficients		$N \times E \times 3$
TER	albedo		$N \times E$
FR	roughness factors		$N \times E$
SM	soil moisture or sea temperature		$N \times E$
SN	snow depth		$N \times E$
PR	surface pressure (or temperature)		$N \times E$

3. Constants

COR	Coriolis force constants	N
SNL	zenith angle of the sun	N
SOL	number of daylight hours	N
CO2	carbon dioxide content	V
03	ozone content	V
ORO	elevation of earth's surface	$N \times E$

[a] This list is a composite of typical quantities used in several programs. No one problem has used them all.

[b] N and E are the north-south, east-west mesh dimensions, respectively. V is the number of vertical levels.

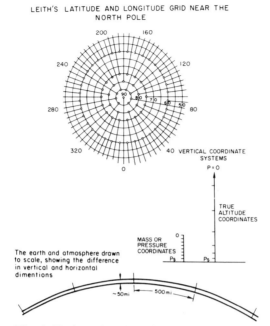

Fig. 4. Horizontal and vertical coordinate systems used in some general circulation calculations.

the poles. Some means of cutting down on the number of computed zones as one approaches the poles are necessary. This creates artificial discontinuities in the mesh that can (and frequently do) cause problems (see Fig. 4).

The second method, which has been used by the National Meteorological Center, the Air Force, and the Fleet Numerical Weather Center, is that of projecting the northern part of the globe onto a plane centered at the North Pole and then subdividing the plane in some regular way. The southern hemisphere, of course, can be done the same way. This method gives the simplest differencing scheme and the best resolution in the upper and middle latitudes. It has difficulties in patching the two hemispheres together. For a detailed discussion of mapping, see Quarles and Spielberg.[31]

The third method is that of using some algorithm for spacing points equally (almost) over the whole globe. Kurihara[32] has described such a system that preserves latitude lines while spacing points almost equally. More future programs probably will use this approach since there is an appreciable storage savings promised.

Table IV. The Number of Mesh Points Required by Numerical Weather Problems as a Function of Resolution

Scale			Number of points			Time step,[c] min	Number of points advanced to compute 1 hr of real time	Typical programs
Spacing (deg latitude)	N^a	Spacing (km at equator)	Number of horizontal mesh points[b]	Number of vertical levels	Total number mesh points			
10	18	1111	648	3	1.9×10^3	20	5.8×10^3	Programs of early 1950's, present streamfunction codes
8	22	889	1,000	2	2.0×10^3	12	1.0×10^4	Mintz's 2-level code
5	36	556	2,592	6	1.5×10^4	10	9.3×10^4	Leith, Smagorinsky's $N = 20$ code
3	60	333	7,200	9	6.5×10^4	6	6.5×10^5	Experimental calculations 1965–1969
2	90	222	16,200	12	1.9×10^5	4	2.9×10^6	Future operational region 1970's
1	180	111	64,800	18	1.2×10^6	2	3.5×10^7	Future experimental region 1970's

[a] N equals twice the number of mesh spaces between the pole and equator.

[b] The number of horizontal points is computed for a 360° × 180° square grid. An actual problem can have 10 to 30% fewer horizontal points, depending on the mapping method used.

[c] The assumption is that each point is advanced every time step. In larger problems, the time step would probably be variable.

5. Time and Space Resolution Problem

In the finite difference methods of solution we have been considering, one of the main characteristics determining the running time of the problem and the accuracy of its results is the number of mesh points used in the calculation. Any of the models is capable of arbitrarily fine refinement in both space and time. Within the limitations of round-off error, the answers will improve the finer the resolution. This creates a certain insatiability in the calculational requirements for such problems. In Table IV we have taken a number of typical mesh sizes that have either been done or discussed as future plans and have computed the total number of spatial mesh points. Since the table was prepared, both the experimental and operational calculations have steadily increased in size. For example, $N = 80$ calculations are now done routinely on an experimental basis.

6. Ratio of the Speed of Calculation to Real Time

The ratio of calculation speed to real time is one of the simplest numbers to state, yet next to the spatial resolution it is the most important in setting the computer requirements. Present models range from 1 to 1, i.e., the calculation proceeding at the same rate as the actual weather to perhaps 10 to 1. Older, greatly simplified models can be integrated on present-day computers at a much higher ratio: 100 to 1 or more.

Table V shows examples of computer speed in terms of millions of instructions per second for different complexity models and different spatial

Table V. Approximate Computer Speeds Required for a Fast Model Computing at 100:1 Times Real Time

Horizontal resolution, spacing in deg latitude	MIPS for various vertical levels				
	3 levels	6 levels	9 levels	12 levels	18 levels
10	0.16	0.32	0.49	0.65	0.97
5	1.3	2.6	3.9	5.2	7.8
3	6.0	12.0	18.0	24.0	36.0
2	20.0	40.0	61.0	81.0	120.0
1	162.0	320.0	490.0	650.0	970.0

A "fast model" is defined as one requiring 1000 computer operations to advance one mesh point one time step on the average.

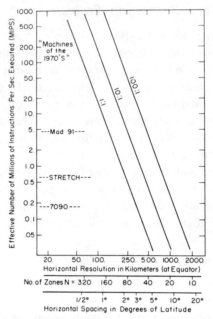

Fig. 5. Computer speed versus horizontal resolution on the globe for a simple model for
various values of $T_{\text{real}}/T_{\text{calculated}}$.

resolutions. Some of these results are shown in the graphs, where they
become straight lines on log-log paper. The graphs are given for particular
combinations of vertical resolution and time step corresponding to diagonal
values in the table (see Fig. 5).

These curves show perhaps better than any other means the true open-
endedness of the computer requirements for general circulation models.
They also show, however, that if one wishes to operate at speeds between
10:1 and 100:1, with a 2° mesh resolution or better with a fairly complicated model, one certainly must get into the range of 20–200 million instructions per second executed.

D. Programming Considerations

1. Complexity Threshold

Personal observations indicate that problem originators each have a
"complexity threshold" beyond which a problem will not be attempted.
This is so not because the problem cannot be solved or cannot be properly

modeled or blocked into arrays, but simply because it has reached a certain point of complexity at which the problem originator decides not to attempt it and will do a simpler job or a different problem instead. This threshold, of course, is quite different for different people and at different times for the same person. However, it is certainly a very important phenomenon, because no matter how many programmers or assistants are put upon a large calculation, it is always a relatively small group, often a single individual—a senior scientist—who really originates the calculation and lays out its overall structure. His threshold of complexity will govern the calculations attempted by his group. If we lower the level of complexity of every single problem by providing a larger random access store, we may extend the frontier of complexity into a whole new range. There is a point where size is no longer simply a question of scaling problems but where it really opens up a whole new class of calculations.

2. Higher-Level Languages and Efficiency

The question of programming system efficiency is very serious for a large problem. The millions of instructions per second used up by an inefficient programming system are just as real as those used on a complicated physical calculation.

It is generally agreed that it is unreasonable to expect people to do large-scale scientific programming in machine language. It is also agreed that they should not be penalized too heavily for using FORTRAN or PL/I. In practice, a combination of FORTRAN supplemented by key data-handling subroutines written in machine language achieves most of the efficiency while retaining the ability to modify and improve the large production problem.

There is now hope that the use of a mathematical programming language, such as APL, can remove some of the arbitrariness from the problem-formulation stage.[33] The purpose of the proposed approach is to eliminate errors in the logical formulation early in the design of a large scientific program. Using a mathematical programming language for expressing the overall program logic in an unambiguous, compact way prevents many of the logical errors that can creep into a program because of its sheer size.

The APL language is preferred here because of the powerful and concise way it represents complex relationships and because it exists in the form of a time-sharing terminal language[34] that enables the testing of parts of the program as they are written.

The following section shows how one proceeds with the analysis and formulation using APL. For more detail see Kolsky.[35]

3. APL Formulation of a Meteorological Problem

The problem of advection in one spatial dimension x is characterized by a fixed fluid velocity u. If Y is the value of a dependent variable, i.e., a quantity embedded in and unchanged at a material point during the flow, we have the equation

$$\frac{DY}{Dt} = \frac{\partial Y}{\partial t} + u\frac{\partial Y}{\partial x} = 0$$

Instead of directly setting up finite-difference approximations of the partial derivatives $\partial/\partial t$ and $\partial/\partial x$, it is helpful to consider the path of a point in the fluid as a function of time. The heavy arrow in Fig. 6 shows the characteristic space-time path of the material point, which is at a position x_i at time t^{n+1}. At time t^n, this point was at position $x_* = u\Delta t$. If we know the value of $Y = Y_*{}^n$ at time t^n, we can set

$$Y_i^{n+1} = Y_*{}^n$$

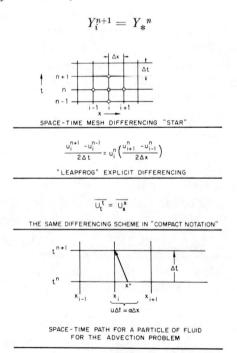

SPACE-TIME MESH DIFFERENCING "STAR"

$$\frac{u_i^{n+1} - u_i^{n-1}}{2\Delta t} = u_i^n\left(\frac{u_{i+1}^n - u_{i-1}^n}{2\Delta x}\right)$$

"LEAPFROG" EXPLICIT DIFFERENCING

$$\overline{u_t^t} = \overline{U_x^x}$$

THE SAME DIFFERENCING SCHEME IN "COMPACT NOTATION"

SPACE-TIME PATH FOR A PARTICLE OF FLUID
FOR THE ADVECTION PROBLEM

Fig. 6. Example of differencing of a partial differential equation for fluid flow.

At time t^n, however, we only know values of Y at mesh points x_{i-1}, x_i, x_{i+1}, etc., and we have to use an interpolation procedure to determine $Y_*{}^n$. Assuming a uniformly spaced mesh of interval Δx, the resulting expression used by Leith[15] is

$$Y_i^{n+1} = Y_*{}^n = Y_i^n - \frac{a}{2}(Y_{i+1}^n - Y_{i-1}^n) + \frac{a^2}{2}(Y_{i+1}^n - 2Y_i^n + Y_{i-1}^n)$$

where

$$a = \frac{x_i - x_*}{\Delta x} = \frac{u\Delta t}{\Delta x}$$

is a dimensionless interpolation parameter. This expression may be written for nonuniform spacing, as shown in the following equation using averaging and differencing operators:

$$Y^* = Y_i + \alpha\left[\frac{Y_i - Y_{i-1}}{a}\frac{b}{a+b} + \frac{Y_{i+1} - Y_i}{b}\frac{a}{a+b}\right]$$
$$+ \alpha^2 \frac{(Y_{i+1} - Y_i)/b - (Y_i - Y_{i-1})/a}{a+b}$$

where $\alpha = x^* - x_i = u\Delta t$, $a = x_i - x_{i-1}$, $b = x_{i+1} - x_i$, and $Y_i^{n+1} = Y_*{}^n$.

One can define a "compact" notation that is an extension of that which has been used by Shuman, Smagorinsky, and others[6,36] in the field. The aim is toward an APL formulation in which the data arrays are considered as a whole and not as isolated components. Define

$$U_t = \frac{u^{t+\frac{1}{2}} - u^{t-\frac{1}{2}}}{\Delta t} \quad \text{and} \quad \overline{U^t} = \frac{u^{t+\frac{1}{2}} + u^{t-\frac{1}{2}}}{2}$$

so that

$$\overline{U_t^t} = \frac{u^{t+1} - u^{t-1}}{2\Delta t}$$

In this notation the subscripts represent differences and the bars represent averages in the variable specified.

The weighted averaging operator is

$$\left(\frac{w}{f^x}\right)_i = A \cdot f_{i+\frac{1}{2}} + B \cdot f_{i-\frac{1}{2}}$$

where

$$A = \frac{\Delta x_{i-\frac{1}{2}}}{\Delta x_{i-\frac{1}{2}} + \Delta x_{i+\frac{1}{2}}} \quad \text{and} \quad B = \frac{\Delta x_{i+\frac{1}{2}}}{\Delta x_{i-\frac{1}{2}} + \Delta x_{i+\frac{1}{2}}}$$

This operator is used particularly in the vertical dimension of the weather model because the vertical dimension is not differenced equally.

The equation for $Y_i{}^*$ may be written in the simpler form

$$Y_i{}^* = Y_i + \alpha\left[\frac{w}{(Y_x)_x}\right]_i + \alpha^2[(Y_x)_x]_i$$

This equation is equivalent to the advection formula, but it is more general because it also applies to nonuniformly spaces meshes. The equation can be written in terms of APL operators for weighted averages and differences as follows:

$$YW \leftarrow YI + (AL1 \times DXAWX\ YI) + AL2 \times DXDX\ YI$$

In APL formulations of problems, it is convenient to define combination operators, such as difference then weighted average (DXAWX), double average (AXAY), and double difference (DXDX). Also, in the case of our meteorological application, ΔY is constant throughout the problem, and ΔX is constant at a given latitude. Thus weighted averages may be replaced by ordinary averages except in the vertical dimension. Combination operators become simpler in such special cases as well as faster to execute on the computer because certain generalized tests are not required.[35]

The real need in a programming system is better coupling between the problem formulators and the computing equipment. The real goal should be to reduce the time from problem formulation to useful answers—not to increase "speed" or "turn-around time" or any of the other usual measures of computing, although they also would improve. Paper improvement in computer efficiency forced by rigid schedules and control program constraints is often exactly the wrong way to measure real progress on a scientific calculation.

E. Future Outlook

Numerical weather prediction has progressed in the last 15 years from theoretical speculations to fully operational networks. The plans for the next ten years, which include the World Weather Watch, automated data collection and communications, promise to outshine the accomplishments of the past.

Concerning the numerical models to be used, the trend toward the finite difference solution of primitive equations will probably continue. Other new methods, such as those based on Fourier transforms, seem to offer little hope. One can expect that more and more detailed physics will be included. Coupled air-sea calculations will be commonly used. Clouds and moisture will be handled much more realistically. The emphasis will probably be on the incorporation of more satellite data, and other exotic measurements, directly into the models.[39,42]

Fortunately the projected computer speeds, storage capacities and data rates for the 1970's seem to match the projected needs in terms of resolution and speed for the global weather problems of the same period. Perhaps this is another case of "feedback" as mentioned earlier, although it will take a tremendous effort from many groups and individuals to bring it to pass.

F. References

1. L. F. Richardson, *Weather Prediction by Numerical Process*, Cambridge University Press, London, 1922 (reprinted by Dover).
2. F. Alt, ed., *Advances in Computers*, Academic Press, New York, 1960, Vol. 1, p. 43 ff.
3. A. Arakawa, "Advanced Topics in Numerical Weather Prediction," Lecture Course Notes prepared by W. E. Langlois, University of California at Los Angeles, 1965.
4. P. D. Thompson, *Numerical Weather Analysis and Prediction*, The Macmillan Company, New York, 1961.
5. National Academy of Sciences, "Weather and Climate Modification—Problems and Prospects," report by Panel on Weather and Climate Modification, *Publ. 1350*, Washington, D.C., 1966.
6. J. Smagorinsky, S. Manabe, and J. L. Holloway, Numerical results from a nine-level general circulation model of the atmosphere, *Monthly Weather Rev.* 93 (12), 727–768 (December 1965); J. L. Holloway and S. Manabe, Simulation of climate by a global general circulation model, *Monthly Weather Rev.* 99 (5), 335 (1971).
7. R. M. Goody, *Atmospheric Radiation*, Oxford Clarendon Press, London, 1964.
8. G. S. Benton, "Interaction Between the Atmosphere and the Oceans," *Publ. 983*, National Academy of Sciences, National Research Council, Washington, D.C., 1962.
9. T. Laevastu, "Synoptic Scale Heat Exchange and Its Relations to Weather," Fleet Numerical Weather Facility, *Tech. Note 7*, 1965.
10. G. P. Cressman, Numerical weather prediction in daily use, *Science* 148, 319–327 (1965).
11. N. A. Phillips, *The Atmosphere and the Sea in Motion*, Oxford Press, London, 1959, pp. 501–504.
12. A. Arakawa, Computational design for long-term numerical integration of the equations of fluid motion: two-dimensional incompressible flow, *J. Comp. Phys.* 1, 119–143 (1966).
13. D. Lilly, On the computational stability of numerical solutions of time-dependent nonlinear geophysical fluid dynamics problems, *Monthly Weather Rev.* 93 (1), 11–26 (1965).
14. P. D. Lax and B. Wendroff, Systems of conservation laws, *Comm. Pure Appl. Math.* 13, 217–237 (1960).

15. C. Leith, "Numerical Simulation of the Earth's Atmosphere," Lawrence Radiation Laboratory, *UCRL* 7986-*T*, 1964.
16. J. G. Charney and N. A. Phillips, Numerical integration of the quasi-geostrophic equations for barotropic and simple baroclinic flows, *J. Meteorol.* **10** (2), 71–99 (April 1953).
17. F. G. Shuman, "Numerical Methods in Weather Prediction," Montly Weather Review **85**, 357–361, also 229–234 (1957); F. G. Shuman and J. B. Hovermale, An operational six-layer primitive equation model, *J. Appl. Meteorol.* **7** (4), 525–547 (August 1968).
18. G. P. Cressman and H. A. Bedient, An experiment in automatic data processing, *Monthly Weather Rev.* **85**, 333–340 (1957).
19. D. Houghton, A. Kasahara, and W. Washington, Long-term integration of the barotropic equations in Eulerian form, *Monthly Weather Rev.* **94** (3) (March 1966); A. Kasahara and W. M. Washington, NCAR global circulation model of the atmosphere, *Monthly Weather Rev.* **95**, 389 (1967); W. M. Washington and A. Kasahara, A January simulation experiment with the two-layer version of the NCAR global circulation model, *Monthly Weather Rev.* **98** (8), p. 559 (1970).
20. Y. Mintz, "Very Long-Term Global Integration of the Primitive Equations of Atmospheric Motion," World Meteorological Organization, *Tech. Note No. 66*, pp. 141–167, 1964; republished as *Am. Meteorol. Soc. Meteorol. Monogr.* **8** (30), (1968).
21. P. M. Wolff, L. P. Carstensen, and T. Laevastu, "Analyses and Forecasting of Sea Surface Temperature," *FNWT Tech. Note No.* 8, 1965.
22. H. M. O'Neil, "The Air Weather Service Six Level Model," *Air Weather Service Tech. Rept. No. 188*, p. 37, Nowember 1966.
23. R. B. Stauffer and T. H. Lewis, MET-WATCH: a technique for processing and scanning meteorological data with a digital computer, *Proc. IFIP Congr. 62, Munich*, pp. 242–246 (1962).
24. J. G. Charney, R. Fjortoft, and J. von Neumann, Numerical integration of the barotropic vorticity equations, *Tellus* **2** (4), 237–254 (1950).
25. J. W. Cooley and J. W. Tukey, An algorithm for the machine calculation of complex Fourier series, *Math. Comp.* **19** (90), 297–301 (April 1965).
26. R. W. Hockney, "A Fast Direct Solution of Poisson's Equation using Fourier Analysis," *Tech. Rept. CS6*, Computer Science Division, Stanford University, April 14, 1964.
27. K. E. Knight, Changes in computer performance, *Datamation* **12** (9), 40 (1966) and **14** (1), 31 (1968).
28. D. Slotnick, W. C. Borck, and R. C. McReynolds, *Proceedings of the Eastern Joint Computer Conference*, Spartan Books, New York, 1962.
29. D. N. Senzig and R. V. Smith, *Proceedings of the Fall Joint Computer Conference*, Spartan Books, New York, 1965.
30. H. G. Kolsky, Some computer aspects of meteorology, *IBM J. Res. Dev.* **11** (6), 584 (1967).
31. D. A. Quarles and K. Spielberg, A computer model for global study of the general circulation of the atmosphere, *IBM J.* **11** (3), 312–336 (1967).
32. Y. Kurihara, Numerical integration of the primitive equations on a spherical grid, *Monthly Weather Rev.* **93** (7), 399 (July 1965).
33. K. E. Iverson, Programming notation in system design, *IBM Syst. J.* **2**, 117–128 (June 1963).

34. *The APL/360 Program, 360D-03.3.007*, which is supported by IBM, and the *APL/360 User's Manual* by A. D. Falkoff and K. E. Iverson (may be obtained through any IBM branch office).

35. H. G. Kolsky, "Problem formulation using APL," *IBM Syst. J.* **8** (3), 240–219 (1969).

36. G. W. Platzman, A retrospective view of Richardson's book on weather prediction, *Bull. Am. Meteorol. Soc.* **48**, 8 (1967).

37. B. H. Armstrong, Theory of the diffusivity factor for atmospheric radiation, *J. Quant. Spectr. Rad. Trans.* **8**, 1577 (1968a); B. H. Armstrong, The radiative diffusivity factor for the random Malkmus band, *J. Atmos. Sci.* **26**, 741 (1969).

38. J. M. Gary, A comparison of difference schemes used for numerical weather prediction, *J. Comp. Phys.* **4** (3), 279 (1969).

39. R. G. Fleagle, "The Atmospheric Sciences and Man's Needs: Priorities for the Future," National Academy of Sciences Committee on Atmospheric Sciences, NAS-NRC, Washington, D.C. (1971).

40. J. E. Fromm, in Frenkiel, *High Speed Computing in Fluid Dynamics*, American Institute of Physics, New York, 1969.

41. L. C. Hobbs, ed., *Parallel Processor Systems, Technologies and Applications*, Spartan and the Macmillan Company, New York, 1970.

42. R. Jastrow and M. Halem, Simulation studies related to GARP, *Bull. Am. Meteorol. Soc.* **51** (6), 490 (1970).

Index